Dénombrement des monocycloalcanes hétéropolysubstitués Tome 1

Alphonse Emadak

Dénombrement des monocycloalcanes hétéropolysubstitués Tome 1

Systèmes ayant une hétéropolysubstitution d'ordre binaire, ternaire, quaternaire, quintuplet et sextuplet

Presses Académiques Francophones

Impressum / Mentions légales

Bibliografische Information der Deutschen Nationalbibliothek: Die Deutsche Nationalbibliothek verzeichnet diese Publikation in der Deutschen Nationalbibliografie; detaillierte bibliografische Daten sind im Internet über http://dnb.d-nb.de abrufbar.

Alle in diesem Buch genannten Marken und Produktnamen unterliegen warenzeichen-, marken- oder patentrechtlichem Schutz bzw. sind Warenzeichen oder eingetragene Warenzeichen der jeweiligen Inhaber. Die Wiedergabe von Marken, Produktnamen, Gebrauchsnamen, Handelsnamen, Warenbezeichnungen u.s.w. in diesem Werk berechtigt auch ohne besondere Kennzeichnung nicht zu der Annahme, dass solche Namen im Sinne der Warenzeichen- und Markenschutzgesetzgebung als frei zu betrachten wären und daher von jedermann benutzt werden dürften.

Information bibliographique publiée par la Deutsche Nationalbibliothek: La Deutsche Nationalbibliothek inscrit cette publication à la Deutsche Nationalbibliografie; des données bibliographiques détaillées sont disponibles sur internet à l'adresse http://dnb.d-nb.de.

Toutes marques et noms de produits mentionnés dans ce livre demeurent sous la protection des marques, des marques déposées et des brevets, et sont des marques ou des marques déposées de leurs détenteurs respectifs. L'utilisation des marques, noms de produits, noms communs, noms commerciaux, descriptions de produits, etc, même sans qu'ils soient mentionnés de façon particulière dans ce livre ne signifie en aucune façon que ces noms peuvent être utilisés sans restriction à l'égard de la législation pour la protection des marques et des marques déposées et pourraient donc être utilisés par quiconque.

Coverbild / Photo de couverture: www.ingimage.com

Verlag / Editeur:
Presses Académiques Francophones
ist ein Imprint der / est une marque déposée de
OmniScriptum GmbH & Co. KG
Heinrich-Böcking-Str. 6-8, 66121 Saarbrücken, Deutschland / Allemagne
Email: info@presses-academiques.com

Herstellung: siehe letzte Seite /
Impression: voir la dernière page
ISBN: 978-3-8381-4002-5

Dénombrement des monocycloalcanes hétéropolysubstitués. Tome 1 : *Systèmes ayant une hétéropolysubstitution d'ordre binaire, ternaire, quaternaire, quintuplet et sextuplet.*

Par

EMADAK Alphonse, PhD

Chargé des Cours à l'Université de Yaoundé I

A Papa Datchoua Pierre

Table des matières

iv

Liste des figures

Liste des tableaux

PRÉFACE

Plusieurs chercheurs ont étudié les problèmes de dénombrement des graphes des stéréoisomères de la série des molécules du cycloalcane monocyclique de formule brute $C_nH_{2n-m}X_m$ où m est le degré de substitution vérifiant la condition $m{\leq}2n$. Les différents auteurs connus ont utilisé soit la méthode de « draw and count », soit le théorème de Polyà pour obtenir une fonction génératrice de dénombrement, soit plus récemment, une correspondance entre théorie de graphe, groupe de symétrie et groupe de permutations pour faire l'inventaire sélectif des nombres de stéréoisomères du cycloalcane homopolysubstitué.

Dans le présent ouvrage, il est essentiellement développé une méthode générale de dénombrement direct des graphes des stéréoisomères d'un cycloalcane monocyclique hétéropolysubstitué de formule brute $C_n X_{m_i} ... Y_{m_i} .. Z_{m_q}$ dans laquelle les entiers positifs n, $m_1,...,m_i,...,m_q$, mis en indice, désignent respectivement la taille du cycle et les degrés de substitution tandis que les termes $X, ..., Y, ..., Z$ symbolisent les substituants de nature distincte qui peuvent être des atomes ou des groupements non isomérisables. Dans cette formule, les super indices $1{\leq}i{\leq}q$ indiquent l'ordre de la substitution.

L'algorithme de dénombrement est construit à partir de la correspondance entre groupe ponctuel de symétrie et groupe de permutations qui permet de générer pour chaque type de permutation, des combinaisons avec répétition des i-uples de substituants $X, ..., Y, ..., Z$ de degrés respectifs $m_1,..., m_i, ..., m_q$, parmi les $2n$ sites de substitution possibles du cycloalcane monocyclique.

En appliquant le lemme de Burnside sur les $4n$ ou les $2n$ opérations du groupe de permutation diédral D_n pour calculer la moyenne de ces opérations appliquées aux permutations dont les longueurs d sont les diviseurs commun de la séquence $(2n, m_1,..., m_i, ..., m_q)$ pour n impair ou $(n, m_1,..., m_i, ..., m_q)$ pour n pair, on détermine les formules de récurrence permettant d'obtenir le nombre de graphes chiraux et achiraux du système hétéropolysubstitué considéré.

En plus de l'introduction générale sur l'historique du dénombrement des isomères et des stéréisomères suivie du chapitre 1 sur l'application de la méthode de correspondance entre groupe ponctuel de symétrie et groupe de permutations pour dénombrer les stéréoisomères du cycloalcane homopolysubstitué, le chapitre 2 comprend entre autres quelques exemples d'application pour les systèmes de tailles $3{\leq}n{\leq}6$, ayant en outre des ordres d'hétéropolysubstitution de nature binaire, ternaire, quaternaire, quintuplet et sextuplet et des degrés variables vérifiant la condition : $\sum_{i=1}^{q} m_i = 2n$. Ces exemples permettront aux étudiants, enseignants, chercheurs ou toutes personnes des domaines de la chimie ou des réseaux numériques d'appréhender et d'appliquer le modèle mathématique établi dans cet ouvrage pour résoudre les problèmes d'hétéropolysubstitution du cycloalcane monocyclique.

Nous exprimons notre gratitude au Prof R. M. Nemba pour avoir guidé nos premiers pas lors de nos travaux sur la topologie moléculaire et la chimie combinatoire énumérative.

Je voudrais remercier ma compagne Claudine Ndioro, mes filles Sana K. Emadak et Doris N. Emadak et la famille Datchoua, pour leur soutien incommensurable pendant la rédaction de cet ouvrage. De même, je remercie l'éditeur ''Presses Académiques Francophones'', Sarrebruck, Allemagne qui a bien voulu publié ce livre.

INTRODUCTION GENERALE

Le problème du dénombrement des isomères s'est posé aux chimistes dés qu'ils ont découvert des composés chimiques ayant la même formule brute et des formules développées distinctes. Les chimistes organiciens tels que Kékulé[1], Couper[1] et Boutlerov[1] vont utiliser la théorie des graphes, découverte par Euler[2] vers 1736, pour effectuer les premiers exercices de dénombrement des isomères de quelques molécules organiques.

En 1857, Cayley [3-4], ayant indépendamment des travaux de Euler[2] découvert la théorie des graphes, va l'associer à la théorie de groupe de permutation publiée par le mathématicien Betti[5-6], pour dénombrer les arborescences. Sa méthode permet de déterminer le nombre des isomères d'alcanes de formule brute (C_nH_{2n+2}) et des radicaux alkyles (C_nH_{2n+1}), de taille n variant de 1 à 13, problème qui était non résolu à cette époque. En outre, ses travaux sur les arborescences, publiés en 1897, marquent le début effectif de l'application de la mathématique à la chimie.

Au début des années 1930, Henze et Blair[7] développent une formule de récurrence qui établit une relation entre le nombre total d'alcools primaires, secondaires et tertiaires pour dénombrer les isomères d'alcools acycliques ($C_nH_{2n+1}OH$) ayant une taille n allant de 1 à 20. Une autre approche de résolution de ce problème proposée par ces derniers consiste à partitionner les alcools en deux sous-classes selon que leur nombre total de carbones est pair ou impair afin d'inventorier les isomères contenus dans chacune d'elles.

Jusqu'au milieu des années 1930, toutes les méthodes de dénombrement des isomères utilisent la représentation géométrique bidimensionnelle des molécules et dénombrent uniquement les isomères relevant de l'isomérie de constitution ou de fonction.

En 1935, Pólya[8] propose dans sa publication intitulée "Combinatorial Enumeration of Groups, Graphs and Chemical Compounds" une méthode de dénombrement des isomères qui utilise le théorème connu sous le nom de "théorème de Pólya" fondé sur le lemme de Burnside[9]. Cette méthode permet de faire un inventaire complet des stéréoisomères des séries de molécules appartenant à une même famille et ayant des degrés de substitution variables. La solution de la méthode d'inventaire de Pólya est une fonction génératrice de la forme $f_i(x) = \sum_i C_i x^i$ où les coefficients C_i sont les nombres de stéréoisomères d'une molécule ayant i substituants.

Depuis lors, des modèles mathématiques de dénombrement des isomères et des stéréoisomères, dérivant ou non du théorème de Pólya, sont proposés et/ou appliqués par plusieurs auteurs, de manière extensive, pour inventorier des familles de molécules polysubstituées. Ainsi, Otter[10] utilise l'équation de dissymétrie des arbres pour dénombrer des types d'arborescences isomorphes. Robinson et al[11] appliquent la méthode de Otter pour déterminer le nombre de stéréoisomères des arborescences quartiques et stériques des alcanes. Balasubramanian[12] emploie le produit de composition ou groupe de Kranz et le produit de composition généralisé pour calculer le nombre des isomères de position des alcanes homopolysubstitués. Balaban[13] utilise le théorème de Pólya pour dénombrer les stéréoisomères des dérivés substitués des familles de molécules telles que les alcanes et les cycloalcanes. Cyvin et al[14] proposent une méthode de dénombrement des structures de résonance ou structures de Kékulé des polyènes conjugués et des polybenzénoïdes. Hässelbarth et al[15] établissent un modèle de classification des mécanismes des réactions de réarrangement. Kerber[16] applique la méthode de coloration pour faire l'inventaire des graphes des molécules organiques et inorganiques. Dolhaine et al[17] mettent au point un algorithme pour le dénombrement des diamutamères et des isomères ayant 2 à plusieurs types de substituants. Plus récemment, Nemba et al[18-23] ont utilisé une correspondance entre groupe de symétrie et groupe de permutations pour faire l'inventaire sélectif des nombres de stéréoisomères du cycloalcane homopolysubstitué de formule brute $C_nH_{2n-m}X_m$ où m est le degré de substitution vérifiant la condition $m \leq 2n$. Les formules de récurrence qu'ils proposent permettent de faire un calcul ponctuel et direct du nombre de stéréoisomères de ce système moléculaire. Cette procédure a l'avantage d'éviter les calculs longs et fastidieux de la méthode de Pólya. Fujita[24-28] propose un modèle de dénombrement par classe de substitution des stéréoisomères des molécules organiques polysubstitués appartenant aux différents sous-groupes de symétrie du groupe de symétrie originel de la molécule parente.

Bien d'autres chercheurs tels que El-Basil[29], Redfield[30], Read[31], De Bruijn[32], Fowler[33], Harary[34], Leroux[35], etc. se sont intéressés à la résolution de la problématique du dénombrement des isomères et des stéréoisomères. Cette revue cependant non exhaustive de scientifiques qui travaillent dans ce champ de recherche démontre son importance et prouve qu'il est en émergence. La raison principale est que l'on découvre plusieurs milliers de molécules chaque année, ce qui induit automatiquement les recherches des formes ayant des propriétés intéressantes pour les applications industrielles que les chercheurs protègent par les brevets et les patentes[36].

Les dérivés substitués des cycloalcanes ont des propriétés recherchées par les chimistes. En effet, les cycloalcanes font partie de la famille des hydrocarbures saturés à chaîne fermée dont la formule chimique générale est C_nH_{2n}. La série déjà mise en évidence par les voies naturelle et/ou de synthèse possède $3 \leq n \leq 288$[37]. On les trouve dans la nature comme composants de certains gisements de pétrole[38] ou des plantes[39]. Toutefois, il est possible de les obtenir par voie de synthèse[40-41]. Ces hydrocarbures ont des propriétés physico-chimiques intéressantes qui sont exploitées non seulement dans l'industrie[41], où ils sont utilisés comme intrants ou intermédiaires dans la fabrication des produits chimiques tels que les solvants, les peintures et les vernis, les fibres de nylon, etc. mais aussi dans la fabrication des médicaments[42].

La détermination du nombre de stéréoisomères de nouveaux dérivés substitués du cycloalcane est donc une première étape indispensable qui est suivie par la

2

constitution, grâce à la chimie combinatoire, des bibliothèques de molécules qui répondent à des besoins industriels ou pharmaceutiques.

L'objectif de cette thèse est de proposer un algorithme de calcul rapide et direct du nombre de graphes des stéréoisomères chiraux et achiraux des molécules du cycloalcane hétéropolysubstitué de formule brute $C_n X_{m_i} ... Y_{m_i} ... Z_{m_q}$ dans laquelle X, ..., Y, ..., Z sont les substituants non isomérisables et de nature distincte. Les entiers positifs n, $m_1, ..., m_i, ..., m_q$ y désignent la taille du cycle ainsi que les degrés de substitution et vérifient la condition : $\sum_{i=1}^{q} m_i = 2n$.

FORMULATION MATHEMATIQUE DU DENOMBREMENT DES GRAPHES DES STEREOISOMERES DU SYSTEME $C_n H_{2n-m} X_m$

1.1. Quelques rappels sur la théorie des graphes

La notion de graphe varie selon qu'on en parle en mathématique ou en chimie. En mathématique, un graphe est un concept abstrait qu'on représente par un ensemble fini d'objets appelés points, reliés entre eux par des lignes (arêtes). Pour les chimistes, un graphe est une structure planaire ou tridimensionnelle formée de points représentant les atomes reliés entre eux par des lignes qui symbolisent les liaisons chimiques.[43-44]

En effet, les chimistes cristallographes ont adopté des figures géométriques, à savoir carré, hexagone, octaèdre, … etc., pour décrire la structure des molécules cristallines. Les chimistes organiciens ont suivi en utilisant des points pour symboliser les atomes de carbone et les lignes pour représenter les liaisons carbone-carbone des composés organiques hydrocarbonés (voir exemple Fig. 1.1 a et c). De même, ils considèrent l'intersection angulaire de deux lignes ainsi que le sommet d'une ligne comme atome de carbone (voir exemple Fig. 1.1 b et d). Certains chimistes représentent les hydrocarbures avec des graphes où l'on omet les hydrogènes et ne tient compte que des atomes de carbone et des liaisons carbone-carbone. Ce genre de représentation plus souple sera utilisé dans ce travail.

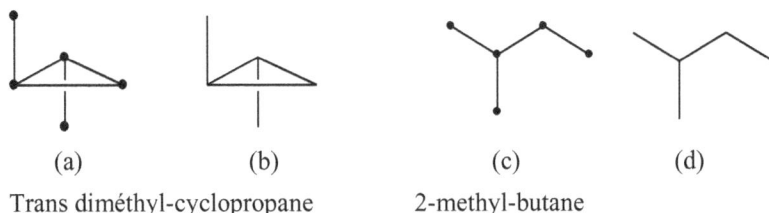

| (a) | (b) | (c) | (d) |

Trans diméthyl-cyclopropane   2-methyl-butane

Figure 1 Exemples de graphes en chimie

Il est opportun de rappeler que dans la théorie des graphes, on ne tient pas compte des paramètres structuraux tels que les angles de valence, de torsion et les longueurs des liaisons.

1.1.1. Correspondance entre théorie des graphes et chimie

En théorie des graphes, *les points* représentent *des atomes*. De même, *les lignes* représentent *les liaisons interatomiques*.

Une molécule est représentée par *un graphe moléculaire* qui est une figure comportant des points et des lignes.

Un graphe moléculaire tridimensionnel ou stéréographe moléculaire est la représentation graphique d'une molécule dans un espace à trois dimensions (Voir Fig. 1a et b).

Un graphe moléculaire bidimensionnel ou *graphe planaire* est la représentation graphique d'une molécule dans un espace à deux dimensions (Voir Fig. 1c et d).

Un graphe planaire cyclique représente une molécule ayant un cycle (Voir Fig. 2a) tandis qu'un graphe planaire acyclique reproduit la forme d'une molécule n'ayant pas de cycle (Voir Fig. 2b).

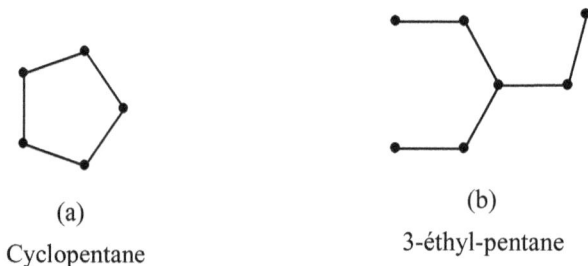

(a)

Cyclopentane

(b)

3-éthyl-pentane

Figure 2 Exemples de graphes cyclique **(a)** et acyclique **(b)**

Tous ces types de graphes sont utilisés pour dénombrer les isomères de constitution et de position ainsi que les stéréoisomères.

1.2. Détermination des permutations engendrées par les opérations de symétrie du groupe D_{nh} agissant sur les $2n$ sites de substitution du stéréographe G d'un cycloalcane

Soit G le stéréographe ou graphe tridimensionnel en symétrie D_{nh} du cycloalcane parent C_nH_{2n}.[21-23, 45-46]

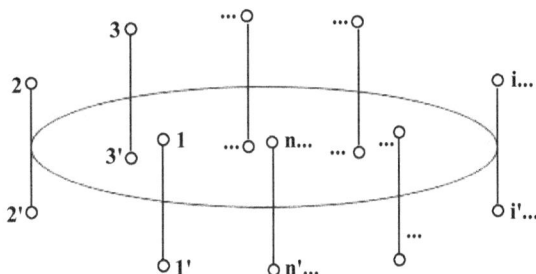

Figure 3 Stéréographe G du cycloalcane parent C_nH_{2n}

Dans G, les positions des 2n hydrogènes numérotées 1, 2, ..., n et 1', 2', ..., n' sont des sites de substitution permutables par les 4n opérations de symétrie de D_{nh} répertoriées dans les expressions suivantes en fonction de la parité de n.

$$D_{nh}=\{E,(n-1)C_n^r,(n-1)S_n^{r'},nC_2,n\sigma_v,\sigma_h\} \qquad \text{si } n \text{ est impair} \qquad (1.1)$$

$$D_{nh} = \{ E,(n-1)C_n^r,(n-2)S_n^{r'},\sigma_h,i,C_2,\frac{n}{2}C',\frac{n}{2}C'',\frac{n}{2}\sigma_v,\frac{n}{2}\sigma_d \} \text{ si } n \text{ est pair} \quad (1.2)$$

Les permutations des $2n$ sites de substitution de G, engendrées par les opérations de symétrie de D_{nh} sont récapitulées dans le tableau 1.[45-46]

Tableau 1 Types de permutations des $2n$ sites de substitution, engendrées par les opérations de symétrie de D_{nh}.

D_{nh}, n impair		D_{nh}, n pair	
Classes d'opérations de symétrie	Permutations engendrées	Classes d'opérations de symétrie	Permutations engendrées
$E = C_n^n$	$[1^{2n}]$	$E = C_n^n$	$[1^{2n}]$
C_n^r r et n premiers entre eux et $r \leq n-1$ $C_n^r = C_{kd}^{kj} = C_d^j$, $j \leq d-1$	$[n^2]$ $\left[d^{\frac{2n}{d}}\right]$	C_n^r r et n premiers entre eux et $r \leq n-1$ $C_n^r = C_{kd}^{kj} = C_d^j$, $j \leq d-1$	$[n^2]$ $\left[d^{\frac{2n}{d}}\right]$
$S_n^{r'}$, r' et n premiers entre eux	$[2n]$	$S_n^{r'}$, r' et n premiers entre eux $S_n^{r'} = S_{kd}^{kj'} = S_d^{j'}$ d pair $j' \leq d-1$	$[n^2]$ $\left[d^{\frac{2n}{d}}\right]$
$S_n^{r'} = S_{kd}^{kj'} = S_d^{j'}$ d impair, $j' \leq 2d-1$	$\left[(2d)^{\frac{n}{d}}\right]$	$S_n^{r'} = S_{kd}^{kj'} = S_d^{j'}$ d impair, $j' \leq 2d-1$	$\left[(2d)^{\frac{n}{d}}\right]$
σ_h	$[2^n]$	σ_h, i, C_2	$[2^n]$, $[2^n]$, $[2^n]$
$n\sigma_v$	$n[1^2 2^{n-1}]$	$\frac{n}{2}\,\sigma_v$	$\frac{n}{2}\,[1^4 2^{n-2}]$
nC_2'	$n[2^n]$	$\frac{n}{2}\,C_2'$, $\frac{n}{2}\,C_2''$, $\frac{n}{2}\,\sigma_d$	$\frac{n}{2}\,[2^n]$, $\frac{n}{2}\,[2^n]$, $\frac{n}{2}\,[2^n]$

Dans le tableau 1 la notation *[iʲ]* dite notation par partition désigne *j* permutations de longueur *i* tandis que les termes *[1²2ⁿ⁻¹]* ou *[1⁴2ⁿ⁻²]* sont les produits des permutations de longueur 1 et 2 lorsque n est impair et pair respectivement.

D'après les données du tableau 1, nous pouvons formuler les remarques suivantes :

Remarque n°1 : les permutations engendrées par les rotoreflexions $S_n^{r'} = \sigma_h\, C_n^{r'}$ et $S_n^{r'} = S_{kd}^{kj'} = S_d^{j'}$ du groupe de symétrie D_{nh} fusionnent les sites de substitution des faces supérieure et inférieure du stéréographe G en une seule permutation de longueur l=2n ou l=2d si n et d sont des nombres impairs ;

Remarque n°2 : les permutations engendrées par les rotoreflexions $S_n^{r'} = \sigma_h \, C_n^{r'}$ et $S_n^{r'} = S_{kd}^{kj'} = S_d^{j'}$ du groupe de symétrie D_{nh} fusionnent en alternant une moitié des sites de la face supérieure et l'autre moitié des sites de la face inférieure pour générer deux permutations distinctes ayant chacune une longueur l=n ou l=d si n ou d sont pairs.

Les différentes permutations des sites de substitutions de G répertoriées dans le tableau 1 peuvent être regroupées dans les ensembles P et P'. L'ensemble P contient les *4n* permutations des sites de substitutions de G engendrées par tous les éléments de symétrie de D_{nh}. Il est défini ainsi qu'il suit:

$$P = \left\{ a_1 \left[1^{2n} \right] (n+1) \left[2^n \right] ..., a_d \left[d^{2n/d} \right] ..., a_n \left[n^2 \right] a_{2n} \left[2n \right] n \left[l^2 2^{n-1} \right] \right\} \quad \text{si } n \text{ est impair} \qquad (1.3)$$

$$P = \left\{ a_1 \left[1^{2n} \right] \frac{3}{2} (n+2) \left[2^n \right] ..., a_d \left[d^{2n/d} \right] ..., a_n \left[n^2 \right] \frac{n}{2} \left[l^4 2^{n-2} \right] \right\} \quad \text{si } n \text{ est pair} \qquad (1.4)$$

Si on supprime dans P les permutations résultant des rotoréflexions et des plans de symétrie, on obtient l'ensemble P' des permutations engendrées par les *2n* opérations de symétrie restantes qui sont l'opération identité E et les rotations propres C_n^r.[45] Ainsi :

$$P' = \left\{ a_1^{/} \left[1^{2n} \right] n \left[2^n \right] ..., a_d^{/} \left[d^{2n/d} \right] ..., a_n^{/} \left[n^2 \right] \right\} \quad \text{si } n \text{ est impair} \qquad (1.5)$$

$$P' = \left\{ a_1^{/} \left[1^{2n} \right], (n+1) \left[2^n \right] ..., a_d^{/} \left[d^{2n/d} \right] ..., a_n^{/} \left[n^2 \right] \right\} \quad \text{si } n \text{ est pair} \qquad (1.6)$$

Dans les expressions précédentes (1.3)-(1.6), les coefficients a_d et $a_d^{/}$ où *$1 \leq d \leq n$ si n est pair* et *$1 \leq d \leq 2n$ si n est impair* représentent les nombres d'opérations de symétrie engendrant les types de permutations *[i^j], [1^2 2^{n-1}]* et *[1^4 2^{n-2}]*.

Pour calculer a_d et $a_d^{/}$, on utilise les relations suivantes :

$a_d = \varphi(d)_{rp}$ d impair premier $\qquad\qquad\qquad\qquad\qquad\qquad (1.7)$

$a_{2d} = \varphi(d)_{ri}$ d impair $\qquad\qquad\qquad\qquad\qquad\qquad\qquad (1.7')$

$a_d = \varphi(d)_{rp} + \varphi(d)_{ri}$ si d(pair parfait ou pair non premier) $\neq 2\theta \, (\theta \text{ impair}) \qquad (1.8)$

$a_d = \varphi(d)_{rp} + \varphi(d)_{ri} + \varphi(\theta)_{ri}$ si d(pair premier) $= 2\theta$ (θ impair) $\qquad (1.9)$

$a_d^{/} = \varphi(d)_{rp}$ si d pair ou impair $\qquad\qquad\qquad\qquad\qquad\qquad (1.10)$

$\varphi(d)_{rp} = \varphi(d)_{ri} \qquad\qquad\qquad\qquad\qquad\qquad\qquad\qquad\qquad (1.11)$

Dans les relations (1.7)-(1.11), les termes $\varphi(d)_{rp}$, $\varphi(d)_{ri}$ et $\varphi(\theta)_{ri}$ sont les fonctions "totient" d'Euler[43] pour les nombres entiers d et θ lorsque l'opération de permutation est équivalente à une rotation propre C_d^j (d porte l'indice rp) ou une rotation impropre $S_d^{j'}$ ou $S_\theta^{j'}$ (d ou θ porte l'indice ri).

Il est à noter que la fonction "totient" de Euler $\varphi(d)$ d'un entier naturel d exprimée ci-après par la relation 1.12 détermine le nombre d'entiers compris entre 1 et d et qui sont premiers avec d.

$$\varphi(d) = d\prod_i\left(1 - \frac{1}{\omega_i}\right) \tag{1.12}$$

où $d \geq 2$ est un entier naturel factorisable sous la forme $d = \omega_1^{\lambda_1}\omega_2^{\lambda_2}..\omega_i^{\lambda_i}..\omega_v^{\lambda_v}$ où ω_i et λ_i sont des entiers positifs non nuls.

Les propriétés de la fonction "totient" d'Euler sont :

a) $\varphi(1) = \varphi(2) = 1$ (1.13)

b) Pour tout entier positif d, $\displaystyle\sum_{\omega_i|d}\varphi(\omega_i) = d$ (1.14)

c) Pour tout entier positif $d = \omega^\lambda$ où ω est premier et λ strictement positif on a :

$$\varphi(d) = \omega^\lambda - \omega^{\lambda-1} \tag{1.15}$$

d) Pour tout entier positif $d = \omega_1\omega_2$ où ω_1 et ω_2 sont des entiers premiers, on a :

$$\varphi(d) = \varphi(\omega_1)\times\varphi(\omega_2) \tag{1.16}$$

e) Pour $d = 2^\lambda$ pair parfait, $\varphi\left(2^\lambda\right) = 2^{\lambda-1}$ (1.17)

f) Pour $d = 2^\lambda\alpha^v$ pair et où α est premier, $\varphi\left(2^\lambda\alpha^v\right) = \varphi\left(2^\lambda\right)\times\varphi\left(\alpha^v\right) = 2^{\lambda-1}\times\left(\alpha^v - \alpha^{v-1}\right)$ (1.18)

En utilisant les propriétés 1.13 à 1.18 dans les relations 1.7 à 1.11 nous obtenons les expressions des coefficients a_d et $a_d^/$ dans le tableau 2 suivant la parité et la divisibilité de d.

Tableau 2 Expressions de a_d et de a'_d suivant la parité et la divisibilité de $d \geq 3$.

Parité de d	d	a_d	a'_d	Domaine de définition
Impair premier	α	$\alpha-1$	$\alpha-1$	$\alpha \geq 3$
Impair non-premier	α^λ	$\alpha^\lambda - \alpha^{\lambda-1}$	$\alpha^\lambda - \alpha^{\lambda-1}$	$\lambda \geq 1$
	$\alpha\beta\gamma...$	$(\alpha-1)(\beta-1)(\gamma-1)...$	$(\alpha-1)(\beta-1)(\gamma-1)...$	$\alpha \geq 3, \beta \geq 3, \gamma \geq 3,...$
	$\alpha^\lambda\beta^\mu$	$(\alpha^\lambda-\alpha^{\lambda-1})(\beta^\mu-\beta^{\mu-1})$	$(\alpha^\lambda-\alpha^{\lambda-1})(\beta^\mu-\beta^{\mu-1})$	$\alpha \geq 3, \beta \geq 3, \lambda \geq 1, \mu \geq 1$
Pair parfait	2^ε	2^ε	$2^{\varepsilon-1}$	$\varepsilon \geq 2$
Pair premier	2α	$3(\alpha-1)$	$(\alpha-1)$	$\alpha > 1$
Pair non-premier	$2^\varepsilon\alpha^\lambda$	$2^\varepsilon(\alpha^\lambda-\alpha^{\lambda-1})$	$2^{\varepsilon-1}(\alpha^\lambda-\alpha^{\lambda-1})$	$\alpha > 1, \varepsilon > 1, \lambda \geq 1$
	$2\alpha^\lambda\beta^\mu$	$2(\alpha^\lambda-\alpha^{\lambda-1})(\beta^\mu-\beta^{\mu-1})$	$(\alpha^\lambda-\alpha^{\lambda-1})(\beta^\mu-\beta^{\mu-1})$	$\alpha > 1, \beta > 1, \lambda \geq 1, \mu \geq 1$
	$2^\varepsilon\alpha^\lambda\beta^\mu$	$2^\varepsilon(\alpha^\lambda-\alpha^{\lambda-1})(\beta^\mu-\beta^{\mu-1})$	$2^{\varepsilon-1}(\alpha^\lambda-\alpha^{\lambda-1})(\beta^\mu-\beta^{\mu-1})$	$\alpha > 1, \beta > 1, \varepsilon > 1, \lambda \geq 1, \mu \geq 1$
	$2\alpha\beta\gamma...$	$3(\alpha-1)(\beta-1)(\gamma-1)...$	$(\alpha-1)(\beta-1)(\gamma-1)...$	$\alpha > 1, \beta > 1, \gamma > 1,...$

La taille n des cycloalcanes déjà mis en évidence par les chimistes varie dans l'intervalle $3 \leq n = d \leq 288$ d'après les données du Chemical Abstracts Service (CAS) Ring System Handbook[37]. Les différentes classes de cycloalcanes découlant de la divisibilité de $3 \leq n \leq 288$ sont reparties ainsi qu'il suit :

1°) Les cycloalcanes dont la taille n n'est pas factorisable en sous-cycles et est un entier premier noté α.

2°) Les cycloalcanes de taille n (impair) factorisable sous la forme $n = \left\{ \begin{array}{l} \alpha^\lambda \\ \alpha^\lambda\beta^\mu \\ \alpha\beta\gamma \end{array} \right\}$,

génèrent des sous-cycles dont la taille d est un facteur ou un produit de facteurs premiers α, β, γ vérifiant les relations ci-avant.

3°) Les cycloalcanes dont la taille n (pair parfait) est factorisable sous la forme $n = 2^\varepsilon$, génèrent des sous-cycles dont la taille $3 \leq d \leq 2^{\varepsilon-1}$ qui sont les diviseurs de n.

4°) Les cycloalcanes dont la taille n (pair) est factorisable sous la forme $n = \left\{ \begin{array}{l} 2^\varepsilon\alpha^\lambda \\ 2^\varepsilon\alpha^\lambda\beta^\mu \\ 2\alpha\beta\gamma \end{array} \right\}$,

génèrent des sous-cycles dont la taille $d \geq 3$ est un diviseur de n ou un produit des diviseurs de n.

Dans ces 4 différentes classes, α, β, γ sont des nombres premiers et ε, λ et μ sont des entiers naturels non nuls.

Exemple 1 : n=15

Posons $n=15=3 \times 5$; les cycles quotients ou sous-cycles ont la taille $d=3$ et 5.

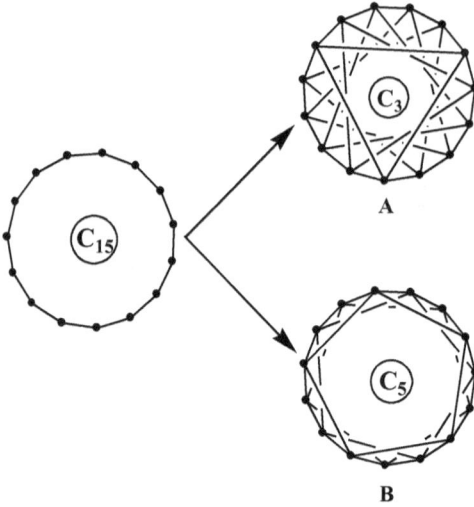

Figure 4 Division d'un cycloalcane de taille $n=15$ en sous-cycles ou cycles quotients de taille $d=3$ (A) et de taille $d=5$ (B).

Exemple 2 : n=18

Posons $n=18=\begin{Bmatrix} 2 \times 3^2 \\ 6 \times 3 \\ 2 \times 9 \end{Bmatrix}$; les cycles quotients ou sous-cycles ont la taille $d=3$, 2×3 et 3^2. On obtient ici des sous-cycles de taille 3, 6 et 9.

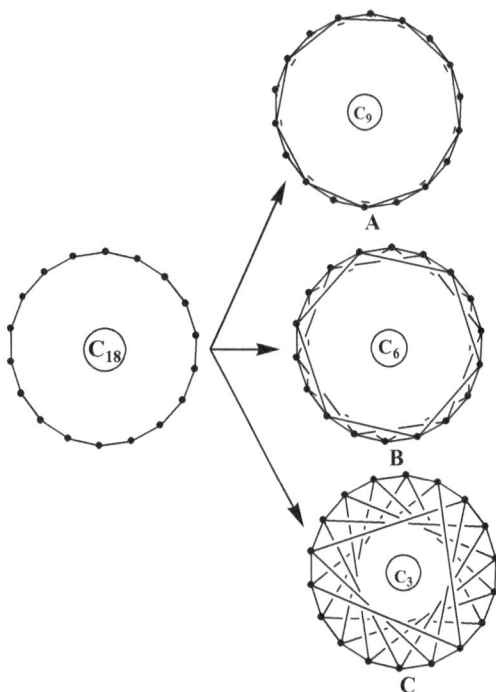

Figure 5 Division d'un cycloalcane de taille $n=18$ en sous-cycles de taille $d=9$ (A), de taille $d=6$ (B) et de taille $d=3$ (C).

La factorisation de n permet de déduire les axes secondaires propres et impropres issus de l'axe principal de rotation C_n. Cette opération permet d'identifier tous les opérateurs de permutation générés par un axe de rotation propre d'ordre n quelconque. Une application faite pour $3 \leq n \leq 288$, nous permet de réaliser la classification des cycloalcanes selon les catégories indiquées dans le tableau 3.

Tableau 3 Table de classification des cycloalcanes

Catégories	I (impair -premier)		II (impair)		III (pair parfait)		IV (pair)	
$n = \prod_i d_i$	α	α^L	$\alpha^L\beta^M$	$\alpha\beta\gamma$	2^K	$2^K\alpha^L$	$2^K\alpha^L\beta^M$	$2\alpha\beta\gamma$
	3	$9=3^2$	$15=3\times5$	$105=3\times5\times7$	$4=2^2$	$6=2\times3$	$30=2\times3\times5$	$210=2\times3\times5\times7$
	5	$25=5^2$	$21=3\times7$	$165=3\times5\times11$	$8=2^3$	$10=2\times5$	$42=2\times3\times7$	
	7	$27=3^3$	$33=3\times11$	$195=3\times5\times13$	$16=2^4$	$12=2^2\times3$	$60=2^2\times3\times5$	
	11	$49=7^2$	$35=5\times7$	$231=3\times7\times11$	$32=2^5$	$14=2\times7$	$66=2\times3\times11$	
	13	$81=3^4$	$39=3\times13$	$255=3\times5\times17$	$64=2^6$	$18=2\times3^2$	$70=2\times5\times7$	
	17	$121=11^2$	$45=3^2\times5$	$285=3\times5\times19$	$128=2^7$	$20=2^2\times5$	$78=2\times3\times13$	
	19	$125=5^3$	$51=3\times17$		$256=2^8$	$22=2\times11$	$84=2^2\times3\times7$	
	23	$169=13^2$	$55=5\times11$			$24=2^3\times3$	$90=2\times3^2\times5$	
	29	$243=3^5$	$57=3\times19$			$26=2\times13$	$102=2\times3\times17$	
	31		$63=3^2\times7$			$28=2^2\times7$	$110=2\times5\times11$	
	37		$65=5\times13$			$34=2\times17$	$114=2\times3\times19$	
	41		$69=3\times23$			$36=2^2\times3^2$	$120=2^3\times3\times5$	
	43		$75=3\times5^2$			$38=2\times19$	$126=2\times3^2\times7$	
	47		$77=7\times11$			$40=2^3\times5$	$130=2\times5\times13$	
	53		$85=5\times17$			$44=2^2\times11$	$132=2^2\times3\times11$	
	59		$87=3\times29$			$46=2\times23$	$138=2\times3\times23$	
	61		$91=7\times13$			$48=2^4\times3$	$140=2^2\times5\times7$	
	67		$93=3\times31$			$50=2\times5^2$	$150=2\times3\times5^2$	
	71		$95=5\times19$			$52=2^2\times13$	$154=2\times7\times11$	
	73		$99=3^2\times11$			$54=2\times3^3$	$156=2^2\times3\times13$	
	79		$111=3\times37$			$56=2^3\times7$	$168=2^3\times3\times7$	
	83		$115=5\times23$			$58=2\times29$	$170=2\times5\times17$	
	89		$117=3^2\times13$			$62=2\times31$	$174=2\times3\times29$	
	97		$119=7\times17$			$68=2^2\times17$	$180=2^2\times3^2\times5$	
	101		$123=3\times41$			$72=2^3\times3^2$	$182=2\times7\times13$	
	103		$129=3\times43$			$74=2\times37$	$186=2\times3\times31$	
	107		$133=7\times19$			$76=2^2\times19$	$190=2\times5\times19$	
	109		$135=3^3\times5$			$80=2^4\times5$	$198=2\times3^2\times11$	
	113		$141=3\times47$			$82=2\times41$	$204=2^2\times3\times17$	
	127		$143=11\times13$			$86=2\times43$	$220=2^2\times5\times11$	
	131		$145=5\times29$			$88=2^3\times11$	$222=2\times3\times37$	
	137		$147=3\times7^2$			$92=2^2\times23$	$228=2^2\times3\times19$	
	139		$153=3^2\times17$			$94=2\times47$	$230=2\times5\times23$	
	149		$155=5\times31$			$96=2^5\times3$	$234=2\times3^2\times13$	
	151		$159=3\times53$			$98=2\times7^2$	$238=2\times7\times17$	
	157		$161=7\times23$			$100=2^2\times5^2$	$240=2^4\times3\times5$	
	163		$171=3^2\times19$			$104=2^3\times13$	$246=2\times3\times41$	
	167		$175=5^2\times7$			$106=2\times53$	$252=2^2\times3^2\times7$	
	173		$177=3\times59$			$108=2^2\times27$	$258=2\times3\times43$	
	179		$183=3\times61$			$112=2^4\times7$	$260=2^2\times5\times13$	
	181		$185=5\times37$			$116=2^2\times29$	$264=2^3\times3\times11$	
	191		$187=11\times17$			$118=2\times59$	$266=2\times7\times19$	
	193		$189=3^3\times7$			$122=2\times61$	$270=2\times3^3\times5$	
	197		$201=3\times67$			$124=2^2\times31$	$276=2^2\times3\times23$	
	199		$203=7\times29$			$134=2\times67$	$280=2^3\times5\times7$	
	211		$205=5\times41$			$136=2^3\times17$	$282=2\times3\times47$	
	223		$207=3^2\times23$			$142=2\times71$	$286=2\times11\times13$	
	227		$209=11\times19$			$144=2^4\times3^2$		
	229		$213=3\times71$			$146=2\times73$		
	233		$215=5\times43$			$148=2^2\times37$		
	239		$217=7\times31$			$152=2^3\times19$		
	241		$219=3\times71$			$158=2\times79$		
	251		$221=13\times17$			$160=2^5\times5$		
	257		$225=3^2\times5^2$			$162=2\times3^4$		
	263		$235=5\times47$			$164=2^2\times41$		
	269		$237=3\times79$			$166=2\times83$		
	271		$245=5\times7^2$			$172=2^2\times43$		
	277		$247=13\times19$			$176=2^4\times11$		
	281		$249=3\times83$			$178=2\times89$		
	283		$253=11\times23$			$184=2^3\times23$		
			$259=7\times37$			$188=2^2\times47$		
			$261=3^2\times29$			$192=2^6\times3$		
			$265=5\times53$			$194=2\times97$		
			$267=3\times89$			$196=2^2\times7^2$		
			$273=3\times91$			$200=2^3\times5^2$		
			$275=5^2\times11$			$202=2\times101$		
			$279=3^2\times31$			$206=2\times103$		
			$287=7\times41$			$208=2^4\times13$		
						$212=2^2\times53$		
						$214=2\times107$		
						$216=2^3\times3^2$		
						$218=2\times109$		
						$224=2^5\times7$		
						$226=2\times113$		
						$232=2^3\times29$		
						$236=2^2\times59$		
						$242=2\times11^2$		
						$244=2^2\times61$		
						$248=2^3\times31$		
						$250=2\times5^3$		
						$254=2\times127$		
						$262=2\times131$		
						$268=2^2\times67$		
						$272=2^4\times17$		
						$274=2\times137$		
						$278=2\times139$		
						$284=2^2\times71$		
						$288=2^5\times3^2$		

1.3. Détermination des permutations des *2n* sites de substitution des systèmes *C$_n$H$_{2n}$* à symétrie *D$_{nh}$* où *n*=3,4,5,6

Dans cette partie, nous appliquons les relations du tableau 1 pour déterminer les types de permutations des 2n sites de substitution engendrées par l'action des opérations de symétrie du groupe *D$_{nh}$* des systèmes *C$_n$H$_{2n}$* où n=3,4,5,6. Nous générons ensuite les tableaux de correspondance entre les notations dites directes, par indice de cycle et par partition de ces permutations.

1.3.1. Cas du Cyclopropane (C$_3$H$_6$)

Considérons le stéréographe *G* du cyclopropane (voir fig. 6 ci-dessous) comme appartenant au groupe ponctuel de symétrie *D$_{3h}$*.

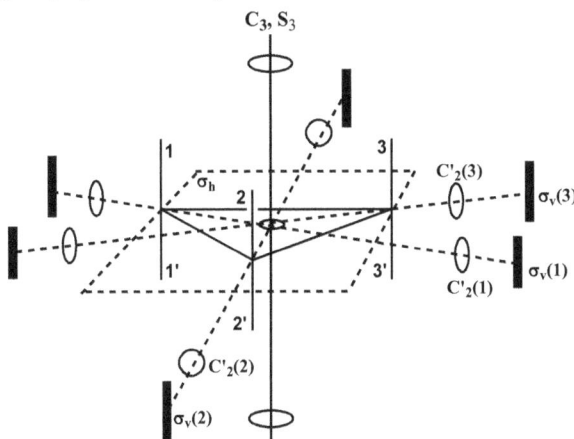

Figure 6 Eléments de symétrie du stéréographe *G* du cyclopropane (*C$_3$H$_6$*) en symétrie *D$_{3h}$*

Les sites de substitution *(SS)* situés sur les faces supérieure et inférieure du graphe *G* du cyclopropane forment l'ensemble *SS* ={1,2,3,1',2',3'}. Les *12* opérations de symétrie de *D$_{3h}$* agissant sur les éléments de l'ensemble *SS* sont répertoriées ainsi qu'il suit : $D_{3h} = \{E, C_3^1, C_3^2, S_3^1, S_3^2, \sigma_h, 3C_2, 3\sigma_v\}$

Les permutations des éléments de l'ensemble *SS* engendrées par les opérations de symétrie de *D$_{3h}$* sont récapitulées dans le tableau 4 selon les notations directes, par partition et par indice de cycle.

14

Tableau 4 Permutations engendrées par les *12* opérations de symétrie de D_{3h} agissant sur les *6* sites de substitution du cyclopropane.

O_{D3h} (G)	Notation directe	Notation par partition	Notation par indice de cycle
E	(1)(2)(3)(1')(2')(3')	$[1^6]$	f_1^6
C_3	(1,2,3)(1',2',3')	$[3^2]$	f_3^2
C_3^2	(1,3,2)(1',3',2')	$[3^2]$	f_3^2
S_3	(1,2',3,1',2,3')	$[6^1]$	f_6^1
S_3^5	(1,3',2,1',3,2')	$[6^1]$	f_6^1
σ_h	(1,1')(2,2')(3,3')	$[2^3]$	f_2^3
$C_2'(1)$	(1,1')(2,3')(2',3)	$[2^3]$	f_2^3
$C_2'(2)$	(2,2')(1,3')(1',3)	$[2^3]$	f_2^3
$C_2'(3)$	(3,3')(1,2')(1',2)	$[2^3]$	f_2^3
$\sigma_v(1)$	(1)(1')(2,3)(2'3')	$[1^2 2^2]$	$f_1^2 f_2^2$
$\sigma_v(2)$	(2)(2')(1,3)(1',3')	$[1^2 2^2]$	$f_1^2 f_2^2$
$\sigma_v(3)$	(3)(3')(2,1)(2',1')	$[1^2 2^2]$	$f_1^2 f_2^2$

D'après la remarque n°1 formulée précédemment, nous constatons que pour $n=3$, les rotoreflexions S_3 et S_3^5 permutent et fusionnent en les alternant, les sites de substitution numérotés *1, 2, 3* de la face supérieure et *1', 2', 3'* de la face inférieure du stéréographe *G* pour engendrer une permutation de longueur *6*. La figure 7 ci-après est une illustration de ce type de permutation engendrée par l'opération de symétrie S_3 agissant sur l'ensemble *SS* de cardinalité 6.

15

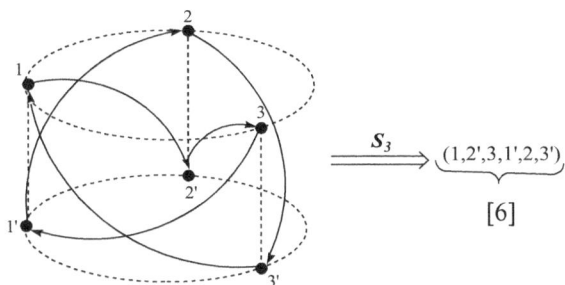

Figure 7 Cycle de permutation de longueur 6 engendré par l'action de S_3 qui fusionne et alterne les 6 sites de substitution du stéréographe G du cyclopropane (C_3H_6.).

N.B. Les flèches indiquent le sens des déplacements alternant des positions des faces inférieure et supérieure.

Les différentes permutations des sites de substitutions de G répertoriées dans le tableau 4 sont regroupées suivant les expressions (1.3) et (1.5) dans les ensembles P et P' ci-après :

P = {$[1^6]$, $2[3^2]$, $2[6^1]$, $4[2^3]$, $3[1^2 2^2]$} ;

P' = {$[1^6]$, $2[3^2]$, $3[2^3]$}.

1.3.2. Cas du Cyclobutane (C_4H_8)

Considérons le stéréographe G du cyclobutane (voir fig. 8 ci-dessous) comme appartenant au groupe ponctuel de symétrie D_{4h}.

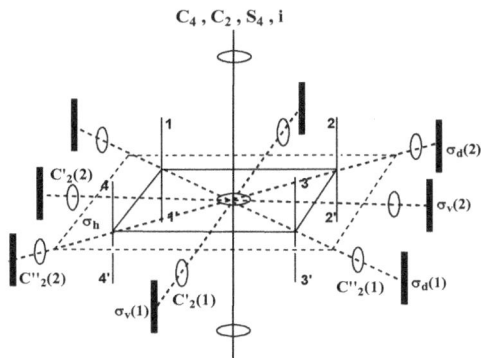

Figure 8 Eléments de symétrie du stéréographe G du cyclobutane (C_4H_8) en symétrie D_{4h}

16

Les sites de substitution du graphe G du cyclobutane ci-dessus forment l'ensemble SS =$\{1,2,3,4,1',2',3',4'\}$. Les 16 opérations de symétrie de D_{4h} agissant sur les éléments de l'ensemble SS sont répertoriées ainsi qu'il suit : $D_{4h} = \{E, C_4^1, C_4^2, C_4^3, S_4^1, S_4^3, i, \sigma_h, 2C_2', 2C_2'', 2\sigma_v, 2\sigma_d\}$

Les permutations des éléments de l'ensemble SS engendrées par les opérations de symétrie de D_{4h} sont récapitulées dans le tableau 5 selon les notations directes, par partition et par indice de cycle.

Tableau 5 Permutations engendrées par les 16 opérations de symétrie de D_{4h} agissant sur les 8 sites de substitutions du cyclobutane.

$O_{D_{4h}}$ (G)	Notation directe	Notation par partition	Notation par indice de cycle
E	(1)(2)(3)(4)(1')(2')(3')(4')	$[1^8]$	f_1^8
C_4	(1,2,3,4)(1',2',3',4')	$[4^2]$	f_4^2
C_4^3	(1,4,3,2)(1',4',3',2')	$[4^2]$	f_4^2
S_4	(1,2',3,4')(1',2,3',4)	$[4^2]$	f_4^2
S_4^3	(1,4',3,2')(1',4,3',2)	$[4^2]$	f_4^2
C_2	(1,3)(2,4)(1',3')(2',4')	$[2^4]$	f_2^4
C_2' (1)	(1,2')(2,1')(3,4')(4,3')	$[2^4]$	f_2^4
C_2' (2)	(1,4')(4,1')(2,3')(3,2')	$[2^4]$	f_2^4
C_2'' (1)	(1,3')(3,1')(2,2')(4,4')	$[2^4]$	f_2^4
C_2'' (2)	(2,4')(4,2')(1,1')(3,3')	$[2^4]$	f_2^4
i	(1,3')(2,4')(3,1')(4,2')	$[2^4]$	f_2^4
σ_h	(1,1')(2,2')(3,3')(4,4')	$[2^4]$	f_2^4
$\sigma_v(1)$	(1)(1')(3)(3')(2,4)(2',4')	$[1^4 2^2]$	$f_1^4 f_2^2$
$\sigma_v(2)$	(2)(2')(4)(4')(1,3)(1',3')	$[1^4 2^2]$	$f_1^4 f_2^2$
$\sigma_d(1)$	(1,2)(1',2')(3,4)(3',4')	$[2^4]$	f_2^4
$\sigma_d(2)$	(1,4)(1',4')(2,3)(2',3')	$[2^4]$	f_2^4

Selon la remarque n°2 formulée précédemment, nous constatons que pour $n=4$, les rotoréflexions S_4 et S_4^3 permutent et fusionnent en deux temps, la moitié des sites numérotés $1, 2, 3, 4$ de la face supérieure de G et la moitié de ceux de sa face inférieure numérotés *1'*, *2'*, *3'* et *4'* pour former deux cycles de permutation de longueur *4* chacune. La figure 9 ci-après est une illustration de cette permutation engendrée par l'opération S_4 agissant sur l'ensemble *SS* de cardinalité 8.

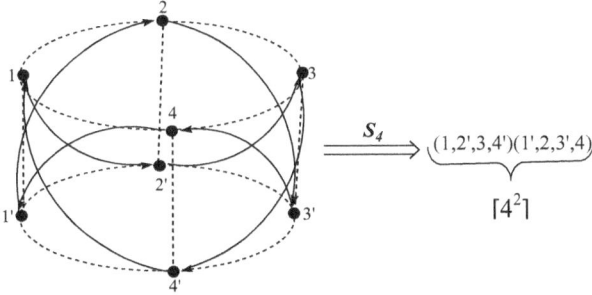

Figure 9 Les 2 cycles de permutations de longueur *d=4* engendrés par l'action de S_4 sur les *8* sites de substitution du stéréographe G du cyclobutane (C_4H_8)

Les différentes permutations des sites de substitutions de *G* répertoriées dans le tableau 5 sont regroupées d'après les expressions (1.4) et (1.6) dans les ensembles *P* et *P'* suivants :

$P = \{[1^8], 4[4^2], 9[2^4], 2[1^4 2^2]\}$;

$P' = \{[1^8], 2[4^2], 5[2^4]\}$.

1.3.3. Cas du Cyclopentane (C_5H_{10})

Considérons le stéréographe *G* du cyclopentane (voir fig. 10) comme appartenant au groupe ponctuel de symétrie D_{5h}.

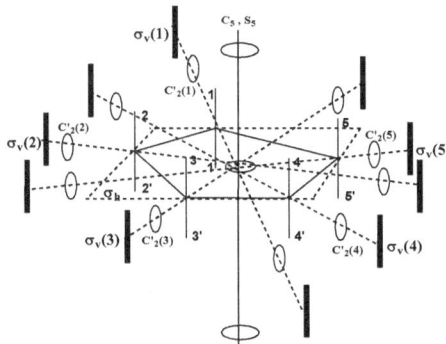

Figure 10 Eléments de symétrie du stéréographe *G* du cyclopentane (C_5H_{10}) en symétrie D_{5h}

18

Les sites de substitution du graphe G du cyclopentane ci-dessus forment l'ensemble $SS=\{1,2,3,4,5,1',2',3',4',5'\}$. Les 20 opérations de symétrie de D_{5h} agissant sur les éléments de l'ensemble SS sont répertoriées ainsi qu'il suit : $D_{5h} = \{E, C_5^1, C_5^2, C_5^3, C_5^4, S_5^1, S_5^2, S_5^3, S_5^4, \sigma_h, 5C_2', 5\sigma_v\}$.

Les permutations des éléments de l'ensemble SS engendrées par les opérations de symétrie de D_{5h} sont récapitulées dans le tableau 6 selon les notations directes, par partition et par indice de cycle.

Tableau 6 Permutations engendrées par les 20 opérations de symétrie de D_{5h} agissant sur les 10 sites de substitution du cyclopentane.

$O_{D_{5h}}$ (G)	Notation directe	Notation par partition	Notation par indice de cycle
E	(1)(2)(3)(4)(5)(1')(2')(3')(4')(5')	$[1^{10}]$	f_1^{10}
C_5	(1,2,3,4,5)(1',2',3',4',5')	$[5^2]$	f_5^2
C_5^2	(1,3,5,2,4)(1',3',5',2',4')	$[5^2]$	f_5^2
C_5^3	(1,4,2,5,3)(1',4',2',5',3')	$[5^2]$	f_5^2
C_4^2	(1,5,4,3,2)(1',5',4',3',2')	$[5^2]$	f_5^2
S_5	(1,2',3,4',5,1',2,3',4,5')	$[10]$	f_{10}^1
S_5^2	(1,3',5,2',4,1',3,5',2,4')	$[10]$	f_{10}^1
S_5^3	(1,4',2,5',3,1',4,2',5,3)	$[10]$	f_{10}^1
S_5^4	(1,5',4,3',2,1',5,4',3,2')	$[10]$	f_{10}^1
$C_2'(1)$	(1,1')(2,5')(3,4')(4,3')(5,2')	$[2^5]$	f_2^5
$C_2'(2)$	(1,3')(2,2')(3,1')(4,5')(5,4')	$[2^5]$	f_2^5
$C_2'(3)$	(1,5')(2,4')(3,3')(4,2')(5,1')	$[2^5]$	f_2^5
$C_2'(4)$	(1,2')(2,1')(3,5')(4,4')(5,3')	$[2^5]$	f_2^5
$C_2'(5)$	(5,5')(1,3')(3,1')(2,4')(4,2')	$[2^5]$	f_2^5
σ_h	(1,1')(2,2')(3,3')(4,4')(5,5')	$[2^5]$	f_2^5
$\sigma_v(1)$	(1)(1')(2,5)(2',5')(3,4)(3',4')	$[1^2 2^4]$	$f_1^2 f_2^4$
$\sigma_v(2)$	(2)(2')(1,3)(1',3')(4,5)(4',5')	$[1^2 2^4]$	$f_1^2 f_2^4$
$\sigma_v(3)$	(3)(3')(2,4)(2',4')(1,5)(1',5')	$[1^2 2^4]$	$f_1^2 f_2^4$
$\sigma_v(4)$	(4)(4')(1,2)(1',2')(3,5)(3',5')	$[1^2 2^4]$	$f_1^2 f_2^4$
$\sigma_v(5)$	(5)(5')(1,4)(1',4')(2,3)(2',3')	$[1^2 2^4]$	$f_1^2 f_2^4$

D'après la remarque n°1 formulée précédemment, nous constatons que pour $n=5$, les rotoreflexions S_5, S_5^2, S_5^3 et S_5^4 permutent et fusionnent en les alternant, les sites de substitution numérotés *1, 2, 3, 4, 5* de la face supérieure et *1', 2', 3', 4', 5'* de la face inférieure du stéréographe *G* pour engendrer une permutation de longueur *10*. La figure 11 ci-après est une illustration de ce type de permutation engendrée par l'opération de symétrie S_5 qui opère une fusion en alternant les sites de substitution placés sur les faces supérieure et inférieure.

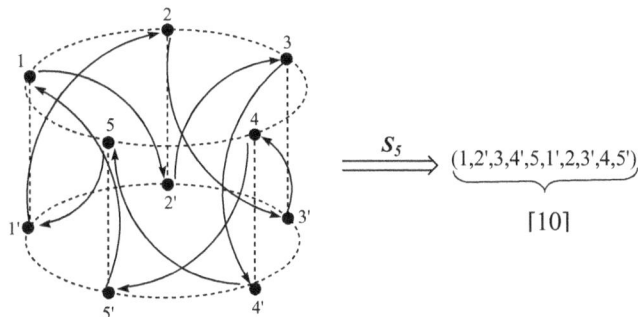

Figure 11 Cycle de permutation de longueur *10* engendré par l'action de S_5 qui fusionne et alterne les *10* sites de substitution du stéréographe G du cyclopentane (C_5H_{10})

Les différentes permutations des sites de substitutions de *G* répertoriées dans le tableau 6 sont regroupées d'après les expressions (1.3) et (1.5) dans les ensembles *P* et *P'*.

P = {$[1^{10}]$, 4$[5^2]$, 4[10], 6$[2^5]$, 5$[1^2 2^4]$};

P' = {$[1^{10}]$, 4$[5^2]$, 5$[2^5]$}.

1.3.4. Cas du Cyclohexane (C_6H_{12})

Considérons le stéréographe *G* du cyclohexane (voir fig. 1.12) comme appartenant au groupe ponctuel de symétrie D_{6h}.

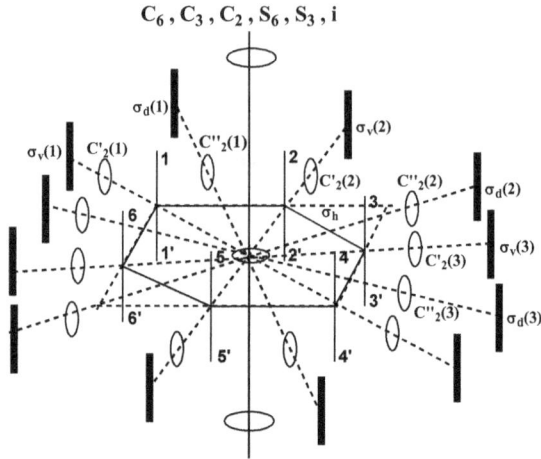

$C_6, C_3, C_2, S_6, S_3, i$

Figure 12 Eléments de symétrie du stéréographe G du cyclohexane C_6H_{12} en symétrie D_{6h}

Les sites de substitution du cyclohexane forment l'ensemble $SS=\{1,2,3,4,5,6,1',2',3',4',5',6'\}$. Les opérations de symétrie de D_{6h} agissant sur les éléments de l'ensemble SS sont : E, C_6^1, $C_6^2=C_3$, $C_6^3=C_2$, $C_6^4=C_3^2$, C_6^5, S_6, S_3, $S_2=i$, S_3^5, S_6^5, σ_h, $3\,C_2'$, $3\,C_2''$, $3\sigma_v$, $3\sigma_d$.

Les permutations des éléments de l'ensemble SS engendrées par les opérations de symétrie de D_{6h} sont récapitulées dans le tableau 7 selon les notations directes, par partition et par indice de cycle.

Tableau 7 Permutations engendrées par les *24* opérations de symétrie de D_{6h} agissant sur les 12 sites de substitution du cyclohexane.

$O_{D_{6h}}$ (G)	Notation directe	Notation par partition	Notation par indice de cycle
E	(1)(2)(3)(4)(5)(6)(1')(2')(3')(4')(5')(6')	$[1^{12}]$	f_1^{12}
C_6	(1,2,3,4,5,6)(1',2',3',4',5',6')	$[6^2]$	f_6^2
$C_6^2 = C_3$	(1,3,5)(2,4,6)(1',3',5')(2',4',6')	$[3^4]$	f_3^4
$C_6^3 = C_2$	(1,4)(2,5)(3,6)(1',4')(2',5')(3',6')	$[2^6]$	f_2^6
$C_6^4 = C_3^2$	(1,5,2)(2,6,3)(1',5',2')(2',6',3')	$[3^4]$	f_3^4
C_6^5	(1,6,5,4,3,2)(1',6',5',4',3',2')	$[6^2]$	f_6^2

21

$O_{D_{6h}}$ (G)	Notation directe	Notation par partition	Notation par indice de cycle
S_6	(1,2',3,4',5,6')(1',2,3',4,5',6)	$[6^2]$	f_6^2
S_3	(1,3',5,1',3,5')(2,4',6,2',4,6')	$[6^2]$	f_6^2
I	(1,4')(1',4)(2,5')(2',5)(3,6')(6',3)	$[2^6]$	f_2^6
S_3^5	(1,5',3,1',5,3')(2,6',4,2',6,4')	$[6^2]$	f_6^2
S_6^5	(1,6',5,4',3,2')(1', 6,5',4,3',2)	$[6^2]$	f_6^2
C_2' (1)	(1,2')(1',2)(3,6')(3',6)(4,5')(4',5)	$[2^6]$	f_2^6
C_2' (2)	(1,4')(1',4)(2,3')(2',3)(5,6')(5',6)	$[2^6]$	f_2^6
C_2' (3)	(1,6')(1',6)(3,4')(3',4)(2,5')(2',5)	$[2^6]$	f_2^6
C_2'' (1)	(1,1')(4,4')(2,6')(2',6)(3,5')(3',5)	$[2^6]$	f_2^6
C_2'' (2)	(2,2')(5,5')(1,3')(1',3)(4,6')(4',6)	$[2^6]$	f_2^6
C_2'' (3)	(3,3')(6,6')(1,5')(1',5)(2,4')(2',4)	$[2^6]$	f_2^6
σ_h	(1,1')(2,2')(3,3')(4,4')(5,5')(6,6')	$[2^6]$	f_2^6
$\sigma_v(1)$	(1,2)(1',2')(3,6)(3',6')(4,5)(4',5')	$[2^6]$	f_2^6
$\sigma_v(2)$	(2,3)(2',3')(1,4)(1',4')(6,5)(6',5')	$[2^6]$	f_2^6
$\sigma_v(3)$	(1,6)(1',6')(3,4)(3',4')(2,5)(2',5')	$[2^6]$	f_2^6
$\sigma_d(1)$	(1)(1')(4)(4')(2,6)(2',6')(3,5)(3',5')	$[1^4 2^4]$	$f_1^4 \, f_2^4$
$\sigma_d(2)$	(2)(2')(5)(5')(1,3)(1',3')(4,6)(4',6')	$[1^4 2^4]$	$f_1^4 \, f_2^4$
$\sigma_d(3)$	(3)(3')(6)(6')(1,5)(1',5')(2,4)(2',4')	$[1^4 2^4]$	$f_1^4 \, f_2^4$

Selon la remarque n°2 formulée précédemment, nous notons d'après le tableau 7 que pour $n=6$, les rotoreflexions permutent et fusionnent en deux temps, la moitié des sites numérotés $1, 2, 3, 4, 5, 6$ de la face supérieure et la moitié de ceux de la face inférieure numérotés $1', 2', 3', 4', 5', 6'$ pour former deux cycles de permutation de longueur 6 chacune.

La figure 13 ci-après est une illustration de cette permutation engendrée par l'opération S_6 agissant sur l'ensemble SS.

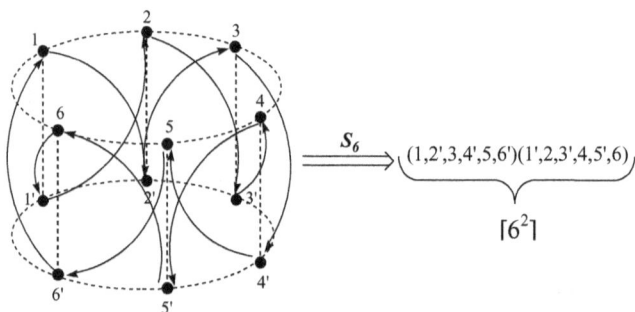

Figure 13 Les *2* cycles de permutations de longueur *d=6* engendrés par l'action de S_6 sur les *12* sites de substitution du stéréographe G du cyclohexane (C_6H_{12})

Les différentes permutations des sites de substitutions de *G* répertoriées dans le tableau 7 sont regroupées d'après les expressions (1.4) et (1.6) dans les ensembles *P* et *P'* suivants :

P = {$[1^{12}]$, $6[6^2]$, $12[2^6]$, $2[3^4]$, $3[1^4 2^4]$};

P' = {$[1^{12}]$, $2[6^2]$, $7[2^6]$, $2[3^4]$}.

Remarque n°3 : Nous considérons que : 1°) Les sites situés sur les faces inférieure et supérieure simulent convenablement les positions endo et exo cycliques des substituants. 2°) Les battements du cycle sont ultra-rapides et dans ces conditions les conformères qui en résulteraient ont une durée de vie très courte.

1.4. Algorithme de dénombrement direct des graphes des stéréoisomères d'un cycloalcane homopolysubstitué

Symbolisons un cycloalcane homopolysubstitué par la formule brute suivante $C_nH_{2n-m}X_m$ dans laquelle les indices *n, m,* et *2n-m* désignent respectivement la taille de la chaîne cyclique carbonée, les nombres de substituants non isomérisables de nature X et d'atomes d'hydrogène H. Dans ce système moléculaire, le couple *(n, m)* vérifie la condition *1≤m≤2n*.

Posons $C_nH_{2n-m}X_m = (n, 2n, m)$. Considérons les ensembles $\mathcal{D}_{2n}$ lorsque *n* est impair, $\mathcal{D}_n$ lorsque *n* est pair et $\mathcal{D}_m$ (*m* pair ou impair), constitués par les éléments qui sont les diviseurs respectifs du triplet de nombres entiers *2n, n* et *m*.

$\mathcal{D}_{2n}=\{1,2,\ldots,d,\ldots n,2n\}$ si n impair

$\mathcal{D}_n=\{1,2,\ldots,d,\ldots,n\}$ si n pair

$\mathcal{D}_m=\{1,\ldots,d',\ldots,m\}$ si *m* pair ou impair

Les éléments des ensembles $\mathcal{D}_{2n}$ et $\mathcal{D}_n$ sont les longueurs des cycles de permutations engendrées par les rotations propres et impropres. Les *2n* sites de substitution du stéréographe G permutables par les opérations de symétrie de D_{nh} obéissent aux règles[3-5] suivantes :

Règle 1.- Pour tout d pair ou impair, chaque opération C_d^r permute en parallèle et de manière disjointe les éléments des faces supérieure et inférieure et engendre (2n/d) d-cycles de permutations notés $\left[d^{\frac{2n}{d}} \right]$ ou $f_d^{\frac{2n}{d}}$.

Règle 2.- Pour tout d impair, l'opération $S_d^{r'}$ permute et fusionne les sites des faces supérieure et inférieure et engendre une permutation de longueur 2d. Pour les 2n sites on obtient n/d cycles de permutations de longueur 2d notés $\left[(2d)^{\frac{n}{d}} \right]$ ou $f_{2d}^{\frac{n}{d}}$.

Si d est pair, l'opération $S_d^{r'}$ permute et fusionne la moitié des sites des faces supérieure et inférieure en une permutation de longueur d. Les 2n sites produiront (2n/d) d-cycles de permutations notés $\left[d^{\frac{2n}{d}} \right]$ ou $f_d^{\frac{2n}{d}}$.

Règle 3.- Toute substitution de degré m sera composée de m/d d-uples de substituants lorsque le nombre entier d est diviseur commun des couples (n,m) ou (2n,m) si n est pair ou impair respectivement.

Règle 4.- Les sous-ensembles de d-cycles de permutations triables dans P et P' et composant une substitution de degré m sont ceux qui vérifient la condition : $d \in \mathcal{D}_{2n} \cap \mathcal{D}_m$ si n est impair ou $d \in \mathcal{D}_n \cap \mathcal{D}_m$ si n est pair.

Pour tout système *(n,2n,m)* où *1≤m≤2n* si les règles 1-4 sont vérifiées les nombre $A_c(n,m)$ de squelettes chiraux et $A_a(n,m)$ de squelettes achiraux sont déterminés à partir des formules de récurrence (1.19)-(1.26) qui sont suivies par des exemples d'applications.[1-2]

Si *n impair, m impair,* $d \neq 2 \in \mathcal{D}_{2n} \cap \mathcal{D}_m$ et $m' = \dfrac{m-1}{2}$, alors :

$$A_c(n,m) = \frac{1}{4n}\left[\sum_{d \neq 2}\left(2a'_d - a_d\right)\binom{\frac{2n}{d}}{\frac{m}{d}} - 2n \cdot \binom{n-1}{m'}\right] \tag{1.19}$$

$$A_a(n,m) = \frac{1}{2n}\left[\sum_{d \neq 2}\left(a_d - a'_d\right)\binom{\frac{2n}{d}}{\frac{m}{d}} + 2n \cdot \binom{n-1}{m'}\right] \tag{1.20}$$

Exemple 1.- n=3, m=3, P={$[1^6]$, $4[2^3]$, $2[3^2]$, $2[6]$, $3[1^2 2^2]$}, P'={$[1^6]$, $3[2^3]$, $2[3^2]$}, $d \in \mathcal{D}_6 \cap \mathcal{D}_3 = \{1,3\}$, $a_1 = a'_1 = 1$, $a_3 = a'_3 = 2$

$$A_c(3,3) = \frac{1}{12}\left[\left((2\times 1) - (1)\right)\binom{6}{3} + (4-2)\binom{2}{1} - (3\times 2)\binom{2}{1}\right] = 1$$

$$A_a(3,3) = \frac{1}{6}\left[\left((1) - (1)\right)\binom{6}{3} + (3\times 2)\binom{2}{1}\right] = 2.$$

Si n impair, m pair, $d \neq 2 \in \mathcal{D}_{2n} \cap \mathcal{D}_m$, alors :

$$A_c(n,m) = \frac{1}{4n}\left[\sum_{d \neq 2}\left(2a'_d - a_d\right)\cdot\binom{\frac{2n}{d}}{\frac{m}{d}} - \binom{n}{\frac{m}{2}}\right] \tag{1.21}$$

$$A_a(n,m) = \frac{1}{2n}\left[\sum_{d \neq 2}\left(a_d - a'_d\right)\binom{\frac{2n}{d}}{\frac{m}{d}} + (n+1)\cdot\binom{n}{\frac{m}{2}}\right] \tag{1.22}$$

Exemple 2.- n=3, m=2, P={$[1^6]$, $4[2^3]$, $2[3^2]$, $2[6]$, $3[1^2 2^2]$}, P'={$[1^6]$, $3[2^3]$, $2[3^2]$}, $d \in \mathcal{D}_6 \cap \mathcal{D}_2 = \{1,2\}$, $a_1 = a'_1 = 1$

$$A_c(3,2) = \frac{1}{12}\left[((2\times 1)-(1))\binom{6}{2} - \binom{3}{1} \right] = 1$$

$$A_a(3,2) = \frac{1}{6}\left[((1)-(1))\binom{6}{2} + (3+1)\binom{3}{1} \right] = 2.$$

Si *n pair, m impair,* $d\neq 2 \in \mathcal{D}_n \cap \mathcal{D}_m$ et $m'=\frac{m-1}{2}$, alors

$$A_c(n,m)=\frac{1}{4n}\left[\sum_{d\neq 2}(2a'_d - a_d)\frac{\binom{2n}{d}}{\frac{m}{d}} - 2n\cdot\binom{n-1}{m'} \right] \tag{1.23}$$

$$A_a(n,m)=\frac{1}{2n}\left[\sum_{d\neq 2}(a_d - a'_d)\frac{\binom{2n}{d}}{\frac{m}{d}} + 2n\cdot\binom{n-1}{m'} \right] \tag{1.24}$$

Exemple 3.- n=4, m=3, P={[1^8], 9[2^4], 4[4^2], 2[1^{4}2^2]}, P'={[1^8], 5[2^4], 2[4^2]}, d∈ $\mathcal{D}_4 \cap \mathcal{D}_3 =\{1\}$, $a_1 = a'_1 = 1$

$$A_c(4,3) = \frac{1}{16}\left[((2\times 1)-(1))\binom{8}{3} - (4\times 2)\binom{3}{1} \right] = 2$$

$$A_a(4,3) = \frac{1}{8}\left[((1)-(1))\binom{8}{3} + (4\times 2)\binom{3}{1} \right] = 3.$$

Si n pair, m pair, d≠2∈ $\mathcal{D}_n \cap \mathcal{D}_m$, alors :

$$A_c(n,m)=\frac{1}{4n}\left[\sum_{d\neq 2}(2a'_d - a_d)\frac{\binom{2n}{d}}{\frac{m}{d}} + \frac{1}{n-1}\left(\frac{m^2}{2}-n(m+1)+1\right)\cdot\binom{n}{\frac{m}{2}} \right] \tag{1.25}$$

$$A_a(n,m)=\frac{1}{2n}\left[\sum_{d\neq 2}(a_d - a'_d)\frac{\binom{2n}{d}}{\frac{m}{d}} + \frac{1}{n-1}\left(n^2+n(m+1)-\frac{m^2}{2}-2\right)\cdot\binom{n}{\frac{m}{2}} \right] \tag{1.26}$$

Exemple 4.- n=4, m=4, P={[1^8], 9[2^4], 4[4^2], 2[1^{4}2^2]}, P'={[1^8], 5[2^4], 2[4^2]}, d∈ $\mathcal{D}_4 \cap \mathcal{D}_4 =\{1,2,4\}$, $a_1 = a'_1 = 1$, $a_4 = 4$, $a'_4 = 2$

$$A_c(4,4) = \frac{1}{16}\left[((2\times 1)-(1))\binom{8}{4} + ((2\times 2)-4)\binom{2}{1} + \frac{1}{3}(\frac{16}{2}-(4\times 5)+1)\binom{4}{2} \right] = 3$$

26

$$A_a(4,4) = \frac{1}{8}\left[((1)-(1))\binom{8}{4} + ((2\times2)-2)\binom{2}{1} + \frac{1}{3}(16+(4\times5)-8-2)\binom{4}{2}\right] = 7.$$

Dans les équations (1.19)-(1.26) la sommation porte sur les éléments $d\neq2$ obéissant à la condition : $d\in\mathcal{D}_{2n}\cap\mathcal{D}_m$ si n est impair ou $d\in\mathcal{D}_n\cap\mathcal{D}_m$ si n est pair ; les quantités a_d et de a'_d sont comme définies précédemment dans les relations (1.7)-(1.10) dans lesquelles les termes de type $\binom{n}{r} = \frac{n!}{r!(n-r)!}$ sont les coefficients binomiaux.

Il est à noter que les degrés de substitution de rang m et $2n$-m sont complémentaires. Cette complémentarité induit les égalités suivantes :

$A_c(n,m) = A_c(n,2n$-$m)$ (1.27)

$A_a(n,m) = A_a(n,2n$-$m)$ (1.27')

qui signifient que les nombres de squelettes chiraux des systèmes $C_nH_{2n\text{-}m}X_m$ et $C_nH_mX_{2n\text{-}m}$ soient respectivement $A_c(n,m)$ et $A_c(n,2n$-$m)$ sont identiques. Cette assertion est également vérifiée pour les squelettes achiraux d'où la relation (1.27').

Ceci est justifiable mathématiquement puisqu'il y a une égalité entre les coefficients multinomiaux suivants :

$$\begin{pmatrix} \dfrac{2n}{d} \\ \dfrac{m}{d} \end{pmatrix} = \begin{pmatrix} \dfrac{2n}{d} \\ \dfrac{2n-m}{d} \end{pmatrix} = \frac{(\frac{2n}{d})!}{(\frac{m}{d})!(\frac{2n-m}{d})!} \qquad (1.28)$$

cette équivalence introduite dans le premier terme des équations (1.19) à (1.26) permet de vérifier les égalités (1.27) et (1.27').

1.5. Programmation des calculs et résultats

Un algorithme de calculs séquentiels a été élaboré et encodé en langage Maple[47] à partir des arguments développés ci-avant.

Les applications pour déterminer les nombres $A_c(n,m)$ de graphes des squelettes chiraux et $A_a(n,m)$ des squelettes achiraux des cycloalcanes homopolysubstitués, $C_nH_{2n\text{-}m}X_m$, où $3\leq n\leq12$ et $1\leq m\leq12$ ont donné les résultats présentés dans le tableau 8.

L'exactitude de ces calculs peut être vérifiée 1°) par le dessin des graphes pour les petites valeurs de n ou 2°) en vérifiant dans la fonction génératrice de Pólya $f_m(x) = \sum C_{m,n}x^m$ d'une série homopolysubstituée que $C_{m,n} = A_c(n,m) + A_a(n,m)$ pour le dénombrement topologique ou $C_{m,n} = 2A_c(n,m) + A_a(n,m)$ pour le dénombrement énantiomérique.[10-11]

1.5.1. *Algorithme séquentiel du dénombrement des graphes des stéréoisomères du cycloalcane homopolysubstitué* $C_nH_{2n-m}X_m$

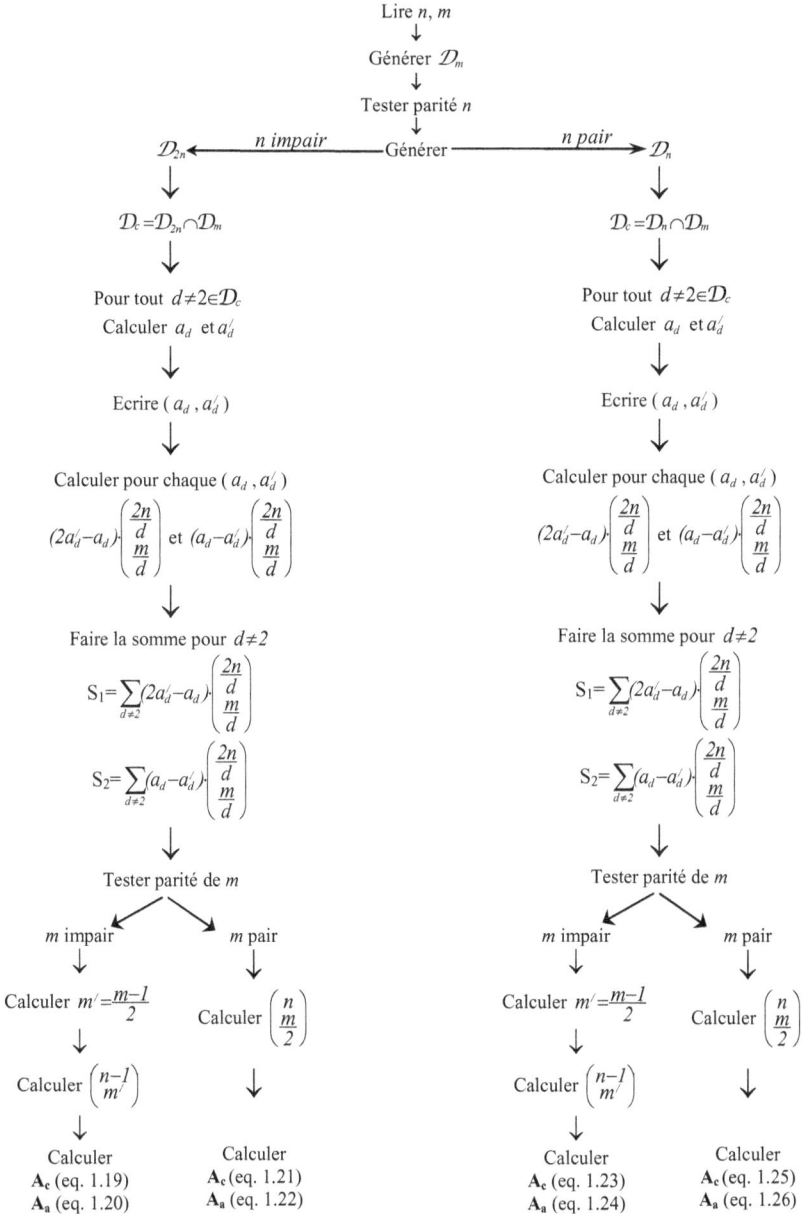

Lire n, m
↓
Générer $\mathcal{D}_m$
↓
Tester parité n
↓

$\mathcal{D}_{2n}$ ←——— *n impair* ——— Générer ——— *n pair* ——→ $\mathcal{D}_n$

↓	↓
$\mathcal{D}_c = \mathcal{D}_{2n} \cap \mathcal{D}_m$	$\mathcal{D}_c = \mathcal{D}_n \cap \mathcal{D}_m$
↓	↓
Pour tout $d \neq 2 \in \mathcal{D}_c$ Calculer a_d et a_d'	Pour tout $d \neq 2 \in \mathcal{D}_c$ Calculer a_d et a_d'
↓	↓
Ecrire (a_d , a_d')	Ecrire (a_d , a_d')
↓	↓
Calculer pour chaque (a_d , a_d') $(2a_d'-a_d)\begin{pmatrix}\frac{2n}{d}\\\frac{m}{d}\end{pmatrix}$ et $(a_d-a_d')\begin{pmatrix}\frac{2n}{d}\\\frac{m}{d}\end{pmatrix}$	Calculer pour chaque (a_d , a_d') $(2a_d'-a_d)\begin{pmatrix}\frac{2n}{d}\\\frac{m}{d}\end{pmatrix}$ et $(a_d-a_d')\begin{pmatrix}\frac{2n}{d}\\\frac{m}{d}\end{pmatrix}$
↓	↓
Faire la somme pour $d \neq 2$ $S_1 = \sum_{d\neq2}(2a_d'-a_d)\begin{pmatrix}\frac{2n}{d}\\\frac{m}{d}\end{pmatrix}$ $S_2 = \sum_{d\neq2}(a_d-a_d')\begin{pmatrix}\frac{2n}{d}\\\frac{m}{d}\end{pmatrix}$	Faire la somme pour $d \neq 2$ $S_1 = \sum_{d\neq2}(2a_d'-a_d)\begin{pmatrix}\frac{2n}{d}\\\frac{m}{d}\end{pmatrix}$ $S_2 = \sum_{d\neq2}(a_d-a_d')\begin{pmatrix}\frac{2n}{d}\\\frac{m}{d}\end{pmatrix}$
↓	↓
Tester parité de m	Tester parité de m

m impair	*m pair*	*m impair*	*m pair*
↓		↓	
Calculer $m' = \frac{m-1}{2}$	Calculer $\begin{pmatrix}n\\\frac{m}{2}\end{pmatrix}$	Calculer $m' = \frac{m-1}{2}$	Calculer $\begin{pmatrix}n\\\frac{m}{2}\end{pmatrix}$
↓	↓	↓	↓
Calculer $\begin{pmatrix}n-1\\m'\end{pmatrix}$		Calculer $\begin{pmatrix}n-1\\m'\end{pmatrix}$	
↓		↓	
Calculer A_c (eq. 1.19) A_a (eq. 1.20)	Calculer A_c (eq. 1.21) A_a (eq. 1.22)	Calculer A_c (eq. 1.23) A_a (eq. 1.24)	Calculer A_c (eq. 1.25) A_a (eq. 1.26)

28

L'encodage de cet algorithme en langage Maple a permis de générer le programme de dénombrement joint en annexe.

Le tableau 8 suivant récapitule le nombre de paires d'énantiomères et de squelettes achiraux des cycloalcanes homopolysubstitués $C_nH_{2n-m}X_m$ où $n=3,4,5,6,7,8,9,10,11,12$ et $1\leq m\leq12$ obtenu à partir de l'exécution du programme de calcul Maple.

Tableau 8 Nombres de paires d'énantiomères et de squelettes achiraux des cycloalcanes homopolysubstitués $C_nH_{2n-m}X_m$ où $n=3,4,5,6,7,8,9,10,11,12$ et $1\leq m\leq12$

	n	3		4		5		6		7	
		A_a	A_c	A_a	A_c	A_a	A_c	A_a	A_c	A_a	A_c
m	0	1	0	1	0	1	0	1	0	1	0
	1	1	0	1	0	1	0	1	0	1	0
	2	2	1	4	1	3	2	5	2	4	3
	3	2	1	3	2	4	4	5	7	6	10
	4			7	3	6	10	14	18	12	35
	5					6	10	10	28	15	64
	6							20	35	20	106
	7									20	113

	n	8		9		10		11		12	
		A_a	A_c	A_a	A_c	A_a	A_c	A_a	A_c	A_a	A_c
m	0	1	0	1	0	1	0	1	0	1	0
	1	1	0	1	0	1	0	1	0	1	0
	2	6	3	5	4	7	4	6	5	8	5
	3	7	14	8	19	9	24	10	30	11	37
	4	24	53	20	84	35	116	30	165	49	215
	5	21	126	28	224	36	370	45	576	55	858
	6	50	241	47	514	100	952	90	1692	174	2778
	7	35	340	56	856	84	1896	120	3816	165	7128
	8	65	390	70	1212	182	3116	180	7260	410	15252
	9			70	1317	126	4136	210	11200	330	27077
	10					222	4578	252	14686	672	40738
	11							252	15907	462	51772
	12									794	56194

1.6. Vérification des résultats

Dans le but de vérifier l'exactitude des calculs nous comparons nos résultats 1°) avec ceux obtenus à partir de la méthode de Pólya[13,18] et 2°) en comptant les graphes dessinés.

1.6.1. Vérification par la méthode de Pólya

Les fonctions de Pólya des dénombrements topologique $F_n^t(x) = \sum_m A^t(n,m)x^m$ et énantiomérique $F_n^e(x) = \sum_m A^e(n,m)x^m$ des graphes des dérivés homopolysubstitués du cyclopropane ($C_3 H_{6-m} (CH_3)_m$) sont données respectivement ci-après :

Pour $n=3$

$$F_3^t(x)=1+x+3x^2+3x^3+3x^4+x^5+x^6$$

$$F_3^e(x)=1+x+4x^2+4x^3+4x^4+x^5+x^6$$

L'inventaire de Pólya des graphes des séries de composés homopolysubstitués du cyclobutane ($C_4 H_{8-m} X_m$), donne :

Pour $n=4$

$$F_4^t(x)=1+x+5x^2+5x^3+10x^4+5x^5+5x^6+x^7+x^8$$

$$F_4^e(x)=1+x+6x^2+7x^3+13x^4+7x^5+6x^6+x^7+x^8$$

On peut déduire par la méthode des différences les valeurs de $A_c(n,m)$ et $A_a(n,m)$ à partir des relations (1.29)-(1.30) suivantes :

$$A_c(n,m)+A_a(n,m)= A^t(n,m) \tag{1.29}$$

$$2A_c(n,m)+A_a(n,m)= A^e(n,m) \tag{1.30}$$

qui lient les coefficients des 2 types de fonctions génératrices.

A titre d'exemple, posons pour le système $C_3H_4X_2$, $(n,m)=(3, 2)$:

$$\begin{cases} A_c(3, 2)+A_a(3, 2)=3 \\ 2A_c(3, 2)+A_a(3, 2)=4 \end{cases} \Longrightarrow \begin{cases} A_c(3, 2)=1 \\ A_a(3, 2)=2 \end{cases}$$

De même, pour le système $C_4H_6X_2$, $(n,m)=(4, 2)$:

$$\begin{cases} A_c(4, 2)+A_a(4, 2)=5 \\ 2A_c(4, 2)+A_a(4, 2)=6 \end{cases} \Longrightarrow \begin{cases} A_c(4, 2)=1 \\ A_a(4, 2)=4. \end{cases}$$

On constate que les valeurs calculées par la méthode des différences tirées des équations ci-dessus et celles obtenues par la méthode combinatoire directe sont identiques à celles du tableau 8 pour l'exemple d'application traité ci-avant.

1.6.2. Vérification par la méthode de comptage des graphes

La deuxième méthode de vérification de la fiabilité de nos calculs c'est de dessiner les graphes de toutes les paires d'énantiomères et des formes achirales et les compter pour obtenir respectivement $A_c(n,m)$ et $A_a(n,m)$. Ceci est possible pour les petites valeurs de n et m.

Nous représentons respectivement dans les Figures 1.14 et 1.15 les graphes moléculaires de toutes les paires d'énantiomères et de tous les squelettes achiraux pour $n=3$ et $1 \leq m \leq 6$ et $n=4$ et $1 \leq m \leq 8$.

1.6.2.1. Paires d'énantiomères et squelettes achiraux du système $C_3 H_{6-m} (CH_3)_m$

m=1: $A_a=1$, $A_c=0$

m=2: $A_a=2$, $A_c=1$

m=3: $A_a=2$, $A_c=1$

m=4: $A_a=2$, $A_c=1$

m=5: $A_a=1$, $A_c=0$

m=6: $A_a=1$, $A_c=0$

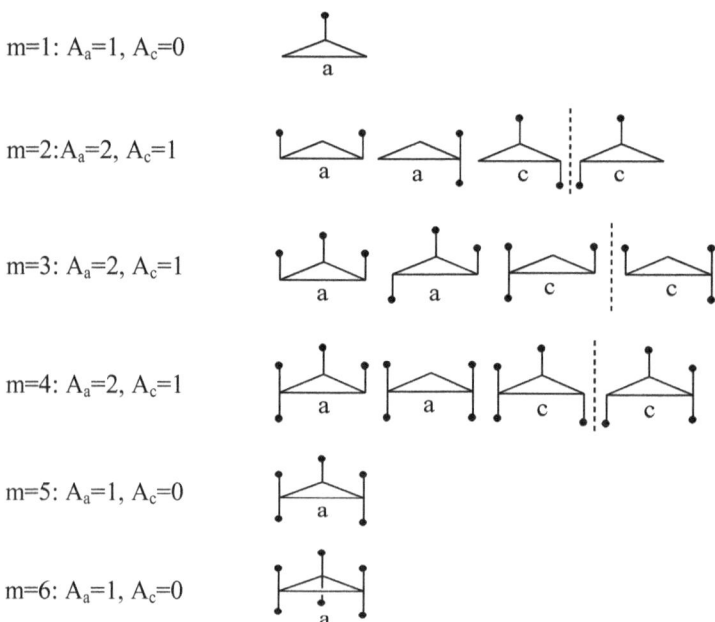

Figure 14 Graphes des paires d'énantiomères et des squelettes achiraux du $C_n H_{2n-m} X_m$ pour $n=3$ et $1 \leq m \leq 6$

Les lettres (a) et (c) placées en dessous des graphes désignent respectivement les isomères achiraux et chiraux tandis que les substituants X sont remplacés par les symboles (●).

31

1.6.2.2. Paires d'énantiomères et squelettes achiraux du système $C_4 H_{8-m} X_m$

m=1: $A_a=1$, $A_c=0$

a

m=2: $A_a=4$, $A_c=1$

m=3: $A_a=3$, $A_c=2$

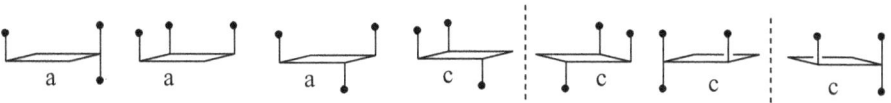

m=4: $A_a=7$, $A_c=3$

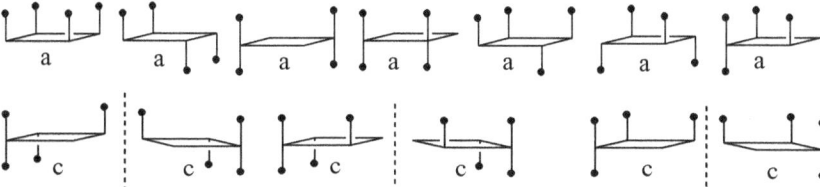

m=5: $A_a=3$, $A_c=2$

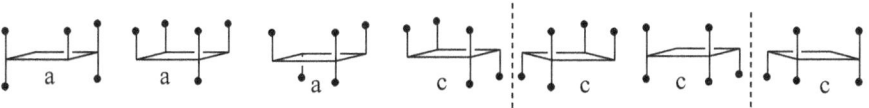

m=6: $A_a=4$, $A_c=1$

m=7: $A_a=1$, $A_c=0$

m=8: $A_a=1$, $A_c=0$

Figure 15 Graphes des paires d'énantiomères et des squelettes achiraux *du système* $C_4 H_{8-m} X_m$ *(1≤m≤8)*

Les lettres (a) et (c) placées en dessous des graphes désignent respectivement les

32

isomères achiraux et chiraux tandis que les substituants X sont remplacés par les symboles (•).

1.7. Exemple de génération des bibliothèques de molécules

En reprenant les graphes dessinés précédemment pour les petits cycles de taille $n=3$ et 4 et en remplaçant (•) par des atomes ou groupes d'atomes non isomérisables tels que Cl, I, F, Br, CH_3, COOH, $COOR_1$, OH, OR_1, COCl, $CONR_1R_2$, C_6H_5, ..., etc. (R_1 et R_2 sont des radicaux alkyles), nous pouvons générer des bibliothèques de molécules du cyclopropane ou du cyclobutane homopolysubstitué.

En remplaçant le substituant (•) des graphes de la figure 14 par le méthyle (CH_3), nous générons la bibliothèque de molécules de la série $C_3 H_{6-m} (CH_3)_m$ ci après illustrée à la figure 16. Certaines molécules connues de cette bibliothèque sont référencées dans la littérature.[48-52] Il s'agit du méthyl cyclopropane (**A**), du cis-diméthyl cyclopropane (**B**), du 1,1-diméthyl cyclopropane (**C**) et du trans-diméthyl cyclopropane (**D**).

De même, en remplaçant le substituant (•) des graphes de la figure 15 par le fluor (F), nous générons la bibliothèque de molécules de la série $C_4 H_{8-m} (F)_m$ dont la plus connue est l'octafluoro cyclobutane. Elle est référencée dans la littérature comme étant un gaz liquéfiable et non inflammable ayant une température d'ébullition de – 5,8°C et une densité de 7,33.[53]

Nous pouvons de la même manière générer autant de bibliothèques de molécules de cycloalcanes homopolysubstitués, $C_n H_{2n-m} X_m$, en faisant varier la taille n du cycle et en choisissant nos substituants X dans une librairie d'atomes et de groupements d'atomes.

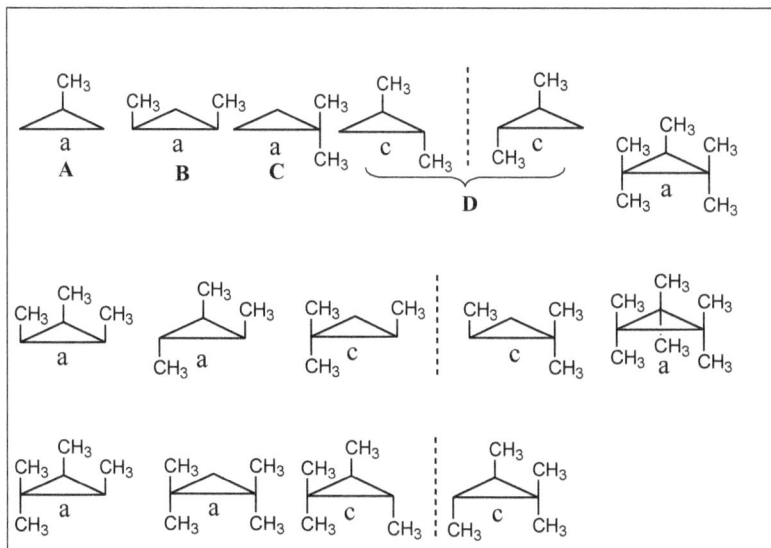

Figure 16 Bibliothèque des molécules du $C_3H_{6-m}(CH_3)_m$

Les lettres (a) et (c) placées en dessous des graphes désignent respectivement les isomères achiraux et chiraux.

CHAPITRE 2

FORMULATION MATHEMATIQUE DU DENOMBREMENT DES GRAPHES DE STEREOISOMERES DU SYSTEME $C_n X_{m_1}...Y_{m_i}...Z_{m_q}$

2.1. Algorithme de dénombrement direct des graphes des stéréoisomères d'un cycloalcane hétéropolysubstitué

Nous développons dans ce chapitre une méthode de détermination directe du nombre de squelettes chiraux et achiraux d'un cycloalcane hétéropolysubstitué ayant une formule brute empirique $C_n X_{m_1}...Y_{m_i}...Z_{m_q}$ et des substituants $X,...,$ Y, ..., Z non isomérisables.

Posons le système $C_n X_{m_1}...Y_{m_i}...Z_{m_q} = (n, m_1,...,m_i...,m_q)$. Les entiers positifs n, $m_1,...,m_i...,m_q$ désignent respectivement la taille du cycle, et les degrés partiels ou nombres d'atomes ou groupes d'atomes de nature X, ..., Y, ... et Z. Le $q+1$-uple de nombres entiers $(n,m_1,...,m_i,...,m_q)$ vérifie la condition : $\sum_{i=1}^{q} m_i = 2n$.

L'ordre de l'hétéropolysubstitution est le nombre de types de substituants non hydrogènes distincts constituant le système $C_n X_{m_1}...Y_{m_i}...Z_{m_q}$.

Une hétéropolysubstitution est d'ordre binaire lorsqu'on a deux types substituants non hydrogènes distincts. Ainsi :

1°) si $q=2$, dans le système $C_n X_{m_1} Y_{m_2}$ dont les degrés partiels vérifient la condition $m_1+m_2=2n$, on a $X \neq Y \neq H$.

2°) si $q=3$, dans le système $C_n X_{m_1} Y_{m_2} Z_{m_3}$ dont les degrés partiels vérifient la condition $m_1+m_2+m_3=2n$, on a $X=H$ et $Y \neq Z \neq H$.

Plus généralement, nous disons qu'une hétéropolysubstitution est d'ordre q-uplet si on a q substituants non hydrogènes distincts dans le système $C_n X_{m_1}...Y_{m_i}...Z_{m_q}$, en d'autres termes $X \neq ... \neq Y \neq ... \neq Z \neq H$.

De même, une hétéropolysubstitution est d'ordre *(q-1)-uplet* si l'on a un atome d'hydrogène parmi les q substituants distincts du système $C_n X_{m_1}...Y_{m_i}...Z_{m_q}$, en d'autres termes $X=H$ et $Y \neq ... \neq V \neq W \neq ... \neq Z \neq H$.

Nous rappelons que la représentation de la molécule parente C_nH_{2n} de la Figure 1 du chapitre 1, est le stéréographe G appartenant au groupe ponctuel de symétrie D_{nh}

d'ordre *4n*, associé au groupe de permutation diédral $\boldsymbol{D_n}$ de même ordre. Soient *P* et *P'* les ensembles des permutations engendrées respectivement par les *4n* et les *2n* opérations de symétrie agissant sur les *2n* sites de substitutions de *G*. *P* et *P'* ont été établis au chapitre 1 suivant les expressions (1.3)-(1.6). Chaque permutation des *2n* sites de substitution de G induite par une opération de symétrie de D_{nh} engendre des combinaisons distinctes avec répétition de $(m_1,...,m_i,...,m_q)$ éléments de différents types *X,...,Y,...* et *Z* parmi *2n* sites de substitution. Ces combinaisons forment des hétéropolysubstitutions distinctes du cycloalcane.

Considérons les ensembles $\mathcal{D}_c = \mathcal{D}_{2n} \cap \mathcal{D}_{m_1} \cap ... \cap \mathcal{D}_{m_i} \cap ... \cap \mathcal{D}_{m_q}$ et $\mathcal{D}_c = \mathcal{D}_n \cap \mathcal{D}_{m_1} \cap ... \cap \mathcal{D}_{m_i} \cap ... \cap \mathcal{D}_{m_q}$ constitués par les éléments qui sont les diviseurs communs respectifs des *(q+1)*-uples de nombres entiers $(2n,m_1,...,m_i,...,m_q) \neq 0$ si *n* est respectivement impair et pair. On obtient pour chaque entier positif $d \in \mathcal{D}_c$ une division des *2n* sites de substitution en $\dfrac{2n}{d}$ boîtes pouvant contenir chacune *d* éléments. On génère ensuite la partition des $(m_1,...,m_i,...,m_q)$ groupements de types *X,...,Y...,Z* en $\left(\dfrac{m_1}{d},...,\dfrac{m_i}{d},...,\dfrac{m_q}{d}\right)$ sous-ensembles de cardinalité *d*. Si chaque élément du *q+1*-uple de nombres $(2n,m_1,...,m_i,...,m_q)$ est divisible par *d*, ou en d'autres termes si *d* est le diviseur commun de $(2n,m_1,...,m_i,...,m_q)$, les propriétés formulées dans les propositions suivantes sont vérifiées.[21-22]

Proposition 1.- Toute hétéropolysubstitution de degrés $(m_1,...,m_i,...,m_q)$ sur le squelette d'un cycloalcane ayant 2n sites de substitution permutables par les 4n opérations du groupe ponctuel de symétrie D_{nh} sera obtenue à partir des arrangements avec répétition de $\left(\dfrac{m_1}{d},...,\dfrac{m_i}{d},...,\dfrac{m_q}{d}\right)$ sous-ensembles de cardinalité d parmi $\dfrac{2n}{d}$ boîtes pouvant contenir chacune d éléments.

Proposition 2.- Seules les permutations du type $\left[d^{\frac{2n}{d}}\right]$ triables dans P et P' et ayant une longueur de cycle égale à d vérifient la proposition 1.

Soit $T(a;b,c,...,e,...,f,g) = \dfrac{a!}{b!\,c!...e!...f!\,g!}$ le coefficient multinomial[43] correspondant au nombre de combinaisons avec répétition de *(b,c,...,e,...,f,g)* objets de nature *X,...,Y,...,Z* parmi *a=2n* boîtes indiscernables tel que $a=b+c+...e+...f+g$.

Les nombres de combinaisons avec répétition résultant des permutations de types $\left[d^{\frac{2n}{d}}\right]$ (où $d \neq 2$), $[2^n]$, $[1^2 2^{n-1}]$ et $[1^4 2^{n-2}]$ répertoriés dans les ensembles *P* et *P'* définis dans le chapitre précédent, sont déterminés conformément aux règles définies par Nemba et al[45-46] ci-après :

Règle 1.- Si $d \in \mathcal{D}_c$, est le diviseur commun de la séquence $(2n, m_1, \ldots, m_i, \ldots, m_q)$ pour n impair ou pair et $\mathcal{D}_c = \{1, \ldots, d, \ldots\}$, alors le nombre de combinaisons distinctes avec répétition ou hétéropolysubstitutions de $(m_1, \ldots, m_i, \ldots, m_q)$ éléments de types X..., Y,... et Z parmi 2n sites de substitution du stéréographe G résultant des permutations $\left[d^{\frac{2n}{d}} \right]$ est donné par le coefficient multinomial :

$$T\left(\frac{2n}{d} ; \frac{m_1}{d}, \ldots, \frac{m_i}{d}, \ldots, \frac{m_q}{d} \right).$$

Règle 2.- Dans le cas des transpositions (2-cyles de permutation ou permutation de longueur d=2) noté $\left[2^n \right]$, le nombre de combinaisons distinctes avec répétition ou hétéropolysubstitutions de $(m_1, \ldots, m_i, \ldots, m_q)$ éléments de types X...,Y,... et Z parmi 2n sites de substitution du stéréographe G résultant de ce type de permutation est :

$$T\left(n ; \frac{m_1}{2}, \ldots, \frac{m_i}{2}, \ldots, \frac{m_q}{2} \right).$$

Règle 3.- Pour déterminer le nombre de combinaisons avec répétition de $(m_1, \ldots, m_i, \ldots, m_q)$ éléments de types distincts X,...,Y,...Z engendrés par les permutations de types $[1^2 2^{n-1}]$ ou $[1^4 2^{n-2}]$ parmi 2n sites de substitution, nous résolvons le système d'équations de partitions complémentaires suivant :

$$l_1 + \ldots + l_i + \ldots + l_q = \begin{cases} 2 & \text{si } n \text{ est impair} \\ \\ 4 & \text{si } n \text{ est pair} \end{cases} \qquad (2.1)$$

$$m_1' + \ldots + m_i' + \ldots + m_q' = \begin{cases} n\text{-}1 & \text{si } n \text{ est impair} \\ \\ n\text{-}2 & \text{si } n \text{ est pair} \end{cases} \qquad (2.2)$$

Dans les équations (2.1) et (2.2), $(l_1, \ldots, l_i, \ldots, l_q) \geq 0$ et $(m_1', \ldots, m_i', \ldots, m_q') \geq 0$ sont les entiers positifs tels que $(l_1, \ldots, l_i, \ldots, l_q)$ indique le choix du placement d'au plus *2* ou *4* éléments de types *X, ..., Y,...Z* sur *2* ou *4* positions invariantes suivant la parité de *n* tandis que $(m_1', \ldots, m_i', \ldots, m_q')$ sont les nombres de couples d'éléments de même nature *X, ..., Y,...* et *Z* à placer parmi *(n-1)* ou *(n-2)* boîtes.

Les couples (l_i, m_i') des solutions $(l_1, ...l_i, ...l_q)$ et $(m_1', ..., m_i', ..., m_q')$ obtenus à partir des équations (2.1) et (2.2) vérifient l'équation (2.3).

$$m_i' = \frac{m_i - l_i}{2} \text{ où } 1 \leq i \leq q \tag{2.3}$$

λ est le nombre de solutions compatibles entre $(l_1, ...l_i, ...l_q)$ et $(m_1, ..., m_i, ..., m_q)$ obtenues à partir des équations (2.1)-(2.2) et vérifiant (2.3).

Le nombre de combinaisons avec répétition de $(m_1, ..., m_i, ..., m_q)$ éléments de types distincts $X, ..., Y, ...Z$ engendrés par les permutations de types $[1^2 2^{n-1}]$ ou $[1^4 2^{n-2}]$ parmi $2n$ sites de substitutions, est déterminé respectivement à partir de la somme sur λ du produit des coefficients multinomiaux par les eq. (2.4) et (2.5) ci-après :

$$\sum_\lambda T(2; l_1, ..., l_i, ..., l_q) \times T(n-1; m_1', ..., m_i', ..., m_q') \qquad \text{si } n \text{ est impair,} \tag{2.4}$$

$$\sum_\lambda T(4; l_1, ..., l_i, ..., l_q) \times T(n-2; m_1', ..., m_i', ..., m_q') \qquad \text{si } n \text{ est pair.} \tag{2.5}$$

Pour toutes les permutations triables dans P et P' obéissant aux propositions et règles ci-dessus énoncées, nous déterminons, suivant les différences $P'-P$ et $2P-P'$ des contributions moyennes des $4n$ permutations de P et $2n$ permutations de P', le couple de nombres entiers $A_c(n, m_1, ..., m_i, ...m_q)$ de graphes chiraux et $A_a(n, m_1, ..., m_i, ...m_q)$ de graphes achiraux du stéréographe G d'un cycloalcane hétéropolysubstitué de la série $C_n X_{m_1} ...Y_{m_i} ...Z_{m_q}$ à partir des formules de récurrence (2.6)-(2.9), suivies par des exemples d'application.

Si n impair :

$$A_c(n, m_1, ..., m_i, ..., m_q) = \frac{1}{4n} \left[\sum_{d \neq 2} (2a_d' - a_d) \cdot \binom{\frac{2n}{d}}{\frac{m_1}{d}, ..., \frac{m_i}{d}, ..., \frac{m_q}{d}} + (n-1) \cdot \binom{n}{\frac{m_1}{2}, ..., \frac{m_i}{2}, ..., \frac{m_q}{2}} \right]$$
$$- \frac{1}{4} \left[\sum_\lambda \binom{2}{l_1, ..., l_i, ..., l_q} \cdot \binom{n-1}{m_1', ..., m_i', ..., m_q'} \right] \tag{2.6}$$

$$A_a(n, m_1, ..., m_i, ..., m_q) = \frac{1}{2n} \left[\sum_{d \neq 2} (a_d - a_d') \cdot \binom{\frac{2n}{d}}{\frac{m_1}{d}, ..., \frac{m_i}{d}, ..., \frac{m_q}{d}} + \binom{n}{\frac{m_1}{2}, ..., \frac{m_i}{2}, ..., \frac{m_q}{2}} \right]$$
$$+ \frac{1}{2} \left[\sum_\lambda \binom{2}{l_1, ..., l_i, ..., l_q} \cdot \binom{n-1}{m_1', ..., m_i', ..., m_q'} \right] \tag{2.7}$$

38

Exemple 1.- Détermination des nombres de graphes chiraux et achiraux du système moléculaire $C_3X_2Y_2Z_2$.

Soit $n = 3, m_1 = 2, m_2 = 2, m_3 = 2$, $\mathcal{D}_6 = \{1,2,3,6\}$

$P = \left\{ [1^6] 4 [2^3] 2 [3^2] 2 [6^1] 3 [1^2 2^4] \right\}$ et $P' = \left\{ [1^6] 3 [2^3] 2 [3^2] \right\}$. L'ensemble des diviseurs communs de la séquence $(6,2,2,2)$ est $\mathcal{D}_c = \{1,2\}$ et $a_1 = a_1' = 1$.

La formule empirique $C_3 X_2 Y_2 Z_2$ contient 3 substituants différents. Les solutions du système de deux équations $l_1 + l_2 + l_3 = 2$ et $m_1' + m_2' + m_3' = 2$ qui peuvent vérifier l'équation (2.3) sont données par chaque ligne des matrices suivantes :

$(l_1, l_2, l_3) = \begin{pmatrix} 200 \\ 020 \\ 002 \end{pmatrix} \rightarrow (m_1', m_2', m_3') = \begin{pmatrix} 011 \\ 101 \\ 110 \end{pmatrix}$. Ainsi $\lambda = 3$ et à partir des équations (2.6)-(2.7) on obtient respectivement :

$$A_c(3,2,2,2) = \frac{1}{12}\left[(2-1)\cdot \begin{pmatrix} 6 \\ 2,2,2 \end{pmatrix} + (3-1)\cdot \begin{pmatrix} 3 \\ \frac{2}{2}, \frac{2}{2}, \frac{2}{2} \end{pmatrix} \right] - \frac{1}{4}\left[\begin{pmatrix} 2 \\ 2,0,0 \end{pmatrix} \cdot \begin{pmatrix} 2 \\ 0,1,1 \end{pmatrix} \right]$$

$$+ \frac{1}{4}\left[\begin{pmatrix} 2 \\ 0,2,0 \end{pmatrix} \cdot \begin{pmatrix} 2 \\ 1,0,1 \end{pmatrix} + \begin{pmatrix} 2 \\ 0,0,2 \end{pmatrix} \cdot \begin{pmatrix} 2 \\ 1,1,0 \end{pmatrix} \right] = 7.$$

$$A_a(3,2,2,2)' = \frac{1}{6}\left[\begin{pmatrix} 3 \\ \frac{2}{2}, \frac{2}{2}, \frac{2}{2} \end{pmatrix} \right] - \frac{1}{4}\left[\begin{pmatrix} 2 \\ 2,0,0 \end{pmatrix} \cdot \begin{pmatrix} 2 \\ 0,1,1 \end{pmatrix} + \begin{pmatrix} 2 \\ 0,2,0 \end{pmatrix} \cdot \begin{pmatrix} 2 \\ 1,0,1 \end{pmatrix} \right]$$

$$+ \frac{1}{4}\left[\begin{pmatrix} 2 \\ 0,0,2 \end{pmatrix} \cdot \begin{pmatrix} 2 \\ 1,1,0 \end{pmatrix} \right] = 4.$$

Exemple 2.- Détermination des nombres de graphes chiraux et achiraux du système moléculaire $C_9 X_9 Y_6 Z_3$.

Soit $n = 9, m_1 = 9, m_2 = 6, m_3 = 3$, $\mathcal{D}_{18} = \{1,2,3,6,9,18\}$,

$P = \left\{ [1^{18}] 10 [2^9] 2 [3^6] 2 [6^3] 6 [9^2] 6 [18] 9 [1^2 2^8] \right\}$ et $P' = \left\{ [1^{18}] 9 [2^9] 2 [3^6] 6 [9^2] \right\}$. L'ensemble des diviseurs communs de la séquence $(18,9,6,3)$ est $\mathcal{D}_c = \{1,3\}$ et $a_1 = a_1' = 1$, $a_3 = a_3' = 2$.

La formule empirique $C_9 X_9 Y_6 Z_3$ contient 3 substituants différents. Les solutions du système de deux équations $l_1 + l_2 + l_3 = 2$ et $m_1' + m_2' + m_3' = 8$ qui peuvent vérifier l'équation (2.3) sont $(l_1, l_2, l_3) = (1,0,1) \rightarrow (m_1', m_2', m_3') = (4,3,1)$. Ainsi, $\lambda = 1$ et à partir des équations (2.6)-(2.7) on obtient respectivement :

$$A_c(9,9,6,3) = \frac{1}{36}\left[(2-1)\cdot\binom{18}{9,6,3} + (4-2)\cdot\begin{pmatrix}\dfrac{18}{3}\\\dfrac{9}{3}\ \dfrac{6}{3}\ \dfrac{3}{3}\end{pmatrix}\right] - \frac{1}{4}\left[\binom{2}{1,0,1}\cdot\binom{8}{4,3,1}\right] = 113310.$$

$$A_a(9,9,6,3) = \frac{1}{18}\left[(9)\cdot\binom{2}{1,0,1}\cdot\binom{8}{4,3,1}\right] = 280.$$

Si n est pair

$$A_c(n,m_1,...,m_i,...,m_q) = \frac{1}{4n}\left[\sum_{d\neq 2}(2a_d' - a_d)\cdot\begin{pmatrix}\dfrac{2n}{d}\\\dfrac{m_1}{d},...,\dfrac{m_i}{d},...,\dfrac{m_q}{d}\end{pmatrix} + \left(\frac{n}{2}-1\right)\cdot\begin{pmatrix}n\\\dfrac{m_1}{2},...,\dfrac{m_i}{2},...,\dfrac{m_q}{2}\end{pmatrix}\right]$$

$$-\frac{1}{8}\left[\sum_{\lambda}\binom{4}{l_1,...,l_i,...,l_q}\cdot\binom{n-2}{m_1',...,m_i',...,m_q'}\right] \qquad (2.8)$$

$$A_a(n,m_1,...,m_i,...,m_q) = \frac{1}{2n}\left[\sum_{d\neq 2}(a_d - a_d')\cdot\begin{pmatrix}\dfrac{2n}{d}\\\dfrac{m_1}{d},...,\dfrac{m_i}{d},...,\dfrac{m_q}{d}\end{pmatrix} + \left(\frac{n}{2}+2\right)\cdot\begin{pmatrix}n\\\dfrac{m_1}{2},...,\dfrac{m_i}{2},...,\dfrac{m_q}{2}\end{pmatrix}\right]$$

$$+\frac{1}{4}\left[\sum_{\lambda}\binom{4}{l_1,...,l_i,...,l_q}\cdot\binom{n-2}{m_1',...,m_i',...,m_q'}\right] \qquad (2.9)$$

Exemple 3.- Détermination des nombres de graphes chiraux et achiraux du système moléculaire $C_6X_6Y_3Z_3$.

Soit $n=6, m_1 = 6, m_2 = 3, m_3 = 3$, $\mathcal{D}_6 = \{1,2,3,6\}$

$P = \{[1^{12}], 12[2^6], 6[6^2], 2[3^4], 3[1^4 2^4]\}$ et $P' = \{[1^{12}], 7[2^6], 2[3^4], 2[6^2]\}$. L'ensemble des diviseurs communs de la séquence $(6,6,3,3)$ est $\mathcal{D}_c = \{1,3\}$ et $a_1 = a_1' = 1, a_3 = a_3' = 2$.

La formule empirique $C_6X_6Y_3Z_3$ contient 3 substituants différents. Les solutions du système de deux équations $l_1 + l_2 + l_3 = 4$ et $m_1' + m_2' + m_3' = 4$ qui peuvent vérifier l'équation (2.3) sont données par chaque ligne des matrices suivantes :

$(l_1,l_2,l_3) = \begin{pmatrix}031\\013\\211\end{pmatrix} \rightarrow (m_1',m_2',m_3') = \begin{pmatrix}301\\310\\211\end{pmatrix}$. Ainsi $\lambda = 3$ et à partir des équations (2.8)-(2.9) on obtient respectivement :

$$A_c(6,6,3,3) = \frac{1}{24}\left[(2-1)\cdot\binom{12}{6,3,3} + (2\times 2-2)\cdot\binom{\frac{12}{3}}{\frac{6}{3},\frac{3}{3},\frac{3}{3}} - 3\binom{4}{0,3,1}\cdot\binom{4}{3,0,1}\right]$$

$$\frac{1}{8}\left[\binom{4}{0,1,3}\cdot\binom{4}{3,1,0} + \binom{4}{2,1,1}\cdot\binom{4}{2,1,1}\right] = 749$$

$$A_a(6,6,3,3) = \frac{1}{12}\left[\frac{1}{8}\left[\binom{4}{0,3,1}\cdot\binom{4}{3,0,1} + \binom{4}{0,1,3}\cdot\binom{4}{3,1,0} + \binom{4}{2,1,1}\cdot\binom{4}{2,1,1}\right]\right] = 44.$$

Exemple 4.- Détermination du nombre de graphes chiraux et achiraux du système moléculaire $C_{12}X_9L_3Y_6Z_6$.

Soit $n=12, m_1=9, m_2=3, m_3=6, m_4=6$, $\mathcal{D}_2=\{1,2,3,4,6,12\}$,

$P=\left\{[1^{24}], 21[2^{12}], 2[3^8], 4[4^6], 6[6^4], 8[12^2], 6[1^4 2^{10}]\right\}$,
$P'=\left\{[1^{24}]13[2^{12}]2[3^8]2[4^6]2[6^4]4[12^2]\right\}$. L'ensemble des diviseurs communs de la séquence $(12, 9, 3, 6)$ est $\mathcal{D}_c=\{1,3\}$, $a_1=a_1'=1$, $a_3=a_3'=2$. La formule empirique $C_{12}X_9L_3Y_6Z_6$ comprend 4 types de substituants différents. Les solutions du système d'équations $l_1+l_2+l_3+l_4=4$ et $m_1'+m_2'+m_3'+m_4'=10$ qui peuvent vérifier l'équation (3.3) sont données par chaque ligne des matrices suivantes :

$$(l_1,l_2,l_3,l_4)=\begin{pmatrix}3100\\1300\\1120\\1102\end{pmatrix}\rightarrow(m_1',m_2',m_3',m_4')=\begin{pmatrix}3133\\4033\\4123\\4132\end{pmatrix}.$$ Ainsi $\lambda=4$ et à partir des équations

(2.8)-(2.9) on obtient respectivement :

$$A_c(12,9,3,6,6) = \frac{1}{48}\left[(2-1)\cdot\binom{24}{9,3,6,6} + (4-2)\cdot\binom{\frac{24}{3}}{\frac{9}{3},\frac{3}{3},\frac{6}{3},\frac{6}{3}}\right]$$

$$-\frac{1}{4}\left[\binom{4}{3,1,0,0}\cdot\binom{10}{3,1,3,3} + \binom{4}{1,3,0,0}\cdot\binom{10}{4,0,3,3}\right. \\ \left. + \binom{4}{1,1,2,0}\cdot\binom{10}{1,1,2,3} + \binom{4}{1,1,0,2}\cdot\binom{10}{4,1,3,2}\right] = 11\,452\,052\,\text{...}$$

$$A_{ac}(12,9,3,6,6) = \frac{1}{4}\left[\begin{pmatrix}4\\3,1,0,0\end{pmatrix}\cdot\begin{pmatrix}10\\3,1,3,3\end{pmatrix} + \begin{pmatrix}4\\1,3,0,0\end{pmatrix}\cdot\begin{pmatrix}10\\4,0,3,3\end{pmatrix} \\ + \begin{pmatrix}4\\1,1,2,0\end{pmatrix}\cdot\begin{pmatrix}10\\1,1,2,3\end{pmatrix} + \begin{pmatrix}4\\1,1,0,2\end{pmatrix}\cdot\begin{pmatrix}10\\4,1,3,2\end{pmatrix}\right] = 96\ 600.$$

2.2. Programmations des calculs et résultats

Un algorithme de calculs séquentiels a été élaboré à partir des arguments développés ci-avant et encodé en langage Maple[47] pour les différents systèmes suivants $C_n X_{m_1} Y_{m_2} Z_{m_3}$, $C_n X_{m_1} Y_{m_2} Z_{m_3} U_{m_4}$, $C_n X_{m_1} Y_{m_2} Z_{m_3} U_{m_4} V_{m_5}$ et $C_n X_{m_1} Y_{m_2} Z_{m_3} U_{m_4} V_{m_5} W_{m_6}$. Les différents programmes de calcul élaborés en langage Maple sont joints en annexe.

Les applications pour déterminer les nombres de squelettes chiraux $A_c(n,m_1,...,m_i,...m_q)$ et de squelettes achiraux $A_a(n,m,...,m_i,...m_q)$ des systèmes $C_n X_{m_1} Y_{m_2} Z_{m_3}$, $C_n X_{m_1} Y_{m_2} Z_{m_3} U_{m_4}$, $C_n X_{m_1} Y_{m_2} Z_{m_3} U_{m_4} V_{m_5}$ et $C_n X_{m_1} Y_{m_2} Z_{m_3} U_{m_4} V_{m_5} W_{m_6}$ où $n = 3,4,5,6$, $m_1+m_2+m_3+...+m_q=2n$ et $3 \leq q \leq 6$ ont donné les résultats présentés dans les tableaux 2.1 à 2.4.

L'exactitude de ces calculs peut être vérifiée 1°) par le dessin des graphes pour les petites valeurs de n ou 2°) en vérifiant dans la fonction génératrice de Pólya

$$f_{m_1,...,m_i,...m_q}(x,...,y,...,z) = \sum_{m_1,...,m_i,...m_q} C_{m_1,...,m_i,...,m_q,n} x^{m_1}...y^{m_i}...z^{m_q}$$ d'une série

polysubstituée que $C_{m_1,...,m_i,...m_q,n} = A_c(n,m_1,...,m_i,...,m_q) + A_a(n, m_1,...,m_i,...,m_q)$ pour le dénombrement topologique ou $C_{m_1,...,m_i,...m_q,n}$ $=2A_c(n,m_1,...,m_i,...,m_q) + A_a(n,m_1,...,m_i,...,m_q)$ pour le dénombrement énantiomérique.

2.2.1. Algorithme séquentiel du dénombrement des graphes des stéréoisomères du cycloalcane hétéropolysubstitué $C_n X_{m_1} ... Y_{m_i} ... Z_{m_q}$

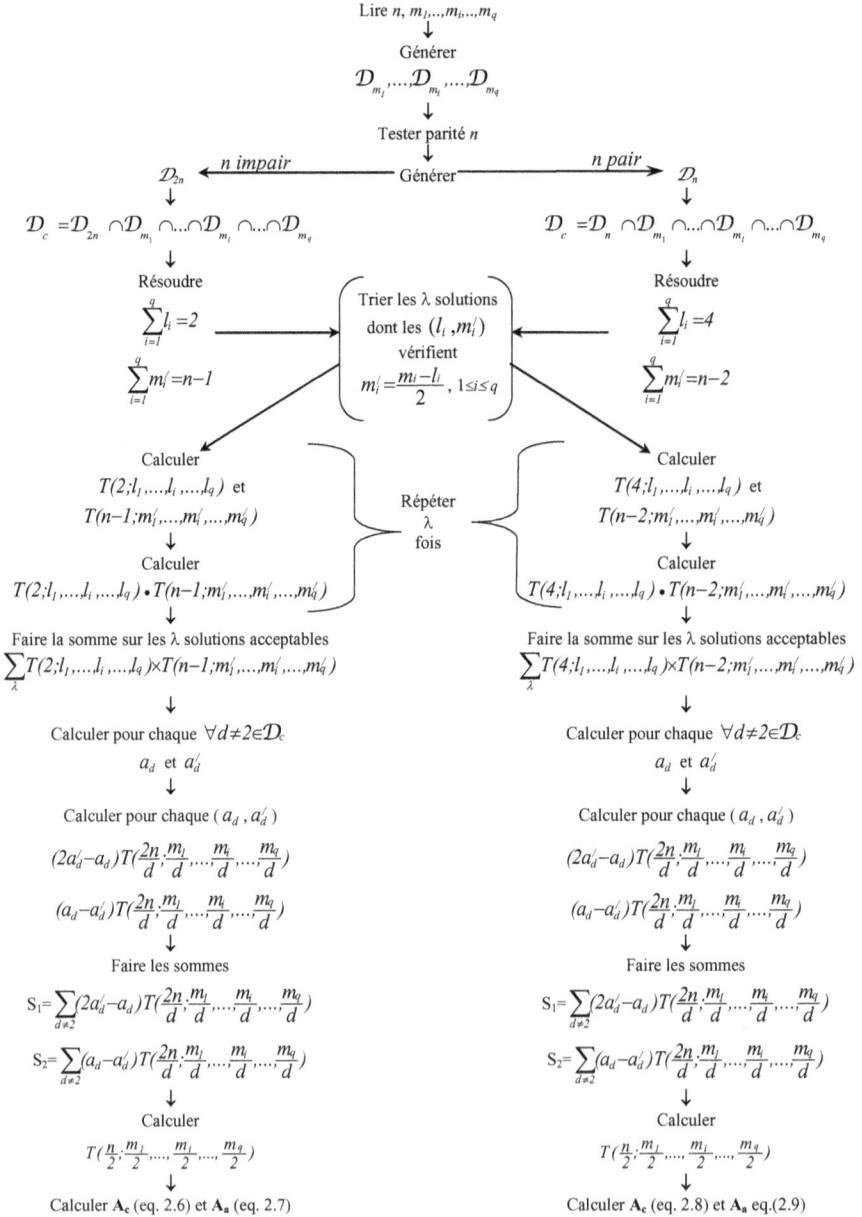

Lire $n, m_1, .., m_i, .., m_q$

↓

Générer

$\mathcal{D}_{m_1}, ..., \mathcal{D}_{m_i}, ..., \mathcal{D}_{m_q}$

↓

Tester parité n

$\xleftarrow{\text{n impair}}$ Générer $\xrightarrow{\text{n pair}}$

$\mathcal{D}_{2n}$ $\mathcal{D}_n$

↓   ↓

$\mathcal{D}_c = \mathcal{D}_{2n} \cap \mathcal{D}_{m_1} \cap ... \cap \mathcal{D}_{m_i} \cap ... \cap \mathcal{D}_{m_q}$ $\mathcal{D}_c = \mathcal{D}_n \cap \mathcal{D}_{m_1} \cap ... \cap \mathcal{D}_{m_i} \cap ... \cap \mathcal{D}_{m_q}$

↓   ↓

Résoudre   Résoudre

$\sum_{i=1}^{q} l_i = 2$ Trier les λ solutions $\sum_{i=1}^{q} l_i = 4$

           dont les (l_i, m_i')

$\sum_{i=1}^{q} m_i' = n-1$ vérifient $\sum_{i=1}^{q} m_i' = n-2$

          $m_i' = \dfrac{m_i - l_i}{2}$, $1 \le i \le q$

Calculer   Répéter   Calculer

$T(2;l_1,...,l_i,...,l_q)$ et λ $T(4;l_1,...,l_i,...,l_q)$ et

$T(n-1;m_1',...,m_i',...,m_q')$ fois $T(n-2;m_1',...,m_i',...,m_q')$

↓   ↓

Calculer   Calculer

$T(2;l_1,...,l_i,...,l_q) \bullet T(n-1;m_1',...,m_i',...,m_q')$ $T(4;l_1,...,l_i,...,l_q) \bullet T(n-2;m_1',...,m_i',...,m_q')$

↓   ↓

Faire la somme sur les λ solutions acceptables   Faire la somme sur les λ solutions acceptables

$\sum_{\lambda} T(2;l_1,...,l_i,...,l_q) \times T(n-1;m_1',...,m_i',...,m_q')$ $\sum_{\lambda} T(4;l_1,...,l_i,...,l_q) \times T(n-2;m_1',...,m_i',...,m_q')$

↓   ↓

Calculer pour chaque $\forall d \ne 2 \in \mathcal{D}_c$ Calculer pour chaque $\forall d \ne 2 \in \mathcal{D}_c$

a_d et a_d' a_d et a_d'

↓   ↓

Calculer pour chaque (a_d, a_d') Calculer pour chaque (a_d, a_d')

$(2a_d' - a_d)T(\dfrac{2n}{d};\dfrac{m_1}{d},...,\dfrac{m_i}{d},...,\dfrac{m_q}{d})$ $(2a_d' - a_d)T(\dfrac{2n}{d};\dfrac{m_1}{d},...,\dfrac{m_i}{d},...,\dfrac{m_q}{d})$

$(a_d - a_d')T(\dfrac{2n}{d};\dfrac{m_1}{d},...,\dfrac{m_i}{d},...,\dfrac{m_q}{d})$ $(a_d - a_d')T(\dfrac{2n}{d};\dfrac{m_1}{d},...,\dfrac{m_i}{d},...,\dfrac{m_q}{d})$

↓   ↓

Faire les sommes   Faire les sommes

$S_1 = \sum_{d \ne 2}(2a_d' - a_d)T(\dfrac{2n}{d};\dfrac{m_1}{d},...,\dfrac{m_i}{d},...,\dfrac{m_q}{d})$ $S_1 = \sum_{d \ne 2}(2a_d' - a_d)T(\dfrac{2n}{d};\dfrac{m_1}{d},...,\dfrac{m_i}{d},...,\dfrac{m_q}{d})$

$S_2 = \sum_{d \ne 2}(a_d - a_d')T(\dfrac{2n}{d};\dfrac{m_1}{d},...,\dfrac{m_i}{d},...,\dfrac{m_q}{d})$ $S_2 = \sum_{d \ne 2}(a_d - a_d')T(\dfrac{2n}{d};\dfrac{m_1}{d},...,\dfrac{m_i}{d},...,\dfrac{m_q}{d})$

↓   ↓

Calculer   Calculer

$T(\dfrac{n}{2};\dfrac{m_1}{2},...,\dfrac{m_i}{2},...,\dfrac{m_q}{2})$ $T(\dfrac{n}{2};\dfrac{m_1}{2},...,\dfrac{m_i}{2},...,\dfrac{m_q}{2})$

↓   ↓

Calculer A_c (eq. 2.6) et A_a (eq. 2.7)   Calculer A_c (eq. 2.8) et A_a eq.(2.9)

Tableau 9 Nombres de paires d'énantiomères et de squelettes achiraux du cycloalcane hétéropolysubstitué de type $C_n X_{m_1} Y_{m_2} Z_{m_3}$ où $n=3,4,5,6$ et $m_1 + m_2 + m_3 = 2n$.

n			3		4		5		6	
m_1	m_2	m_3	A_c	A_a	A_c	A_a	A_c	A_a	A_c	A_a
4	1	1	2	1						
3	2	1	4	2						
2	2	2	7	4						
6	1	1			2	3				
5	2	1			8	5				
4	2	2			23	14				
4	3	1			14	7				
3	3	2			30	10				
8	1	1					4	1		
7	2	1					16	4		
6	3	1					40	4		
6	2	2					62	12		
5	4	1					60	6		
5	3	2					120	12		
4	4	2					156	18		
4	3	3					204	12		
10	1	1							4	3
9	2	1							24	7
8	3	1							76	13
8	2	2							118	29
7	4	1							156	18
7	3	2							316	28
6	5	1							220	22
6	4	2							564	62
6	3	3							749	44
5	5	2							672	42
5	4	3							1128	54
4	4	4							1422	96

Tableau 10 Nombres de paires d'énantiomères et de squelettes achiraux du cycloalcane hétéropolysubstitué de type $C_n X_{m_1} Y_{m_2} Z_{m_3} U_{m_4}$ où $n=3,4,5,6$ et $m_1+m_2+m_3+m_4=2n$.

n				3		4		5		6	
m_1	m_2	m_3	m_4	A_c	A_a	A_c	A_a	A_c	A_a	A_c	A_a
3	1	1	1	10	0						
2	1	2	1	14	2						
5	1	1	1			18	6				
4	1	2	1			48	9				
3	1	3	1			64	12				
3	2	2	1			98	14				
2	2	2	2			150	30				
7	1	1	1					36	0		
6	1	2	1					124	4		
5	1	3	1					252	0		
5	2	2	1					372	12		
4	1	4	1					312	6		
4	2	3	1					624	12		
3	3	3	1					840	0		
3	2	3	2					1248	24		
9	1	1	1							52	6
8	1	2	1							240	15
7	1	3	1							648	24
7	2	2	1							972	36
6	1	4	1							1140	30
6	2	3	1							2284	52
6	2	2	2							3436	128
5	1	5	1							1368	36
5	2	4	1							3432	66
5	3	3	1							4584	72
5	2	3	2							6876	108
4	2	4	2							8616	198
4	3	3	2							11484	132
3	3	3	3							15330	144

45

Tableau 11 Nombres de paires d'énantiomères et de squelettes achiraux du cycloalcane hétéropolysubstitué de type $C_n X_{m_1} Y_{m_2} Z_{m_3} U_{m_4} V_{m_5}$ où $n=3,4,5,6$ et $m_1+m_2+m_3+m_4+m_5=2n$.

n					3		4		5		6	
m_1	m_2	m_3	m_4	m_5	A_c	A_a	A_c	A_a	A_c	A_a	A_c	A_a
2	1	1	1	1	30	0						
4	1	1	1	1			102	6				
3	1	1	2	1			204	12				
6	1	1	1	1					252	0		
5	1	1	2	1					756	0		
4	1	1	3	1					1260	0		
4	1	2	2	1					1884	12		
3	1	2	3	1					2520	0		
3	2	2	2	1					3768	24		
2	2	2	2	2					5664	72		
8	1	1	1	1							492	6
7	1	1	2	1							1968	24
6	1	1	3	1							4608	24
6	1	2	2	1							6900	60
5	1	1	4	1							6912	36
5	1	2	3	1							13824	72
5	2	2	2	1							20724	132
4	1	2	4	1							17280	90
4	1	3	3	1							23064	72
4	2	2	3	1							34572	156
4	2	2	2	2							51897	366
3	2	2	3	2							69168	264

Tableau 12 Nombres de paires d'énantiomères et de squelettes achiraux du cycloalcane hétéropolysubstitué de type $C_n X_{m_1} Y_{m_2} Z_{m_3} U_{m_4} V_{m_5} W_{m_6}$ où $n=3,4,5,6$ et $m_1+m_2+m_3+m_4+m_5+m_6=2n$.

m_1	m_2	m_3	m_4	m_5	m_6	n=3 A_c / A_a		n=4 A_c / A_a		n=5 A_c / A_a		n=6 A_c / A_a	
1	1	1	1	1	1	60	0						
3	1	1	1	1	1			420	0				
2	1	1	1	2	1			624	12				
5	1	1	1	1	1					1512	0		
4	1	1	1	2	1					3780	0		
3	1	1	1	3	1					5040	0		
3	1	1	2	2	1					7560	0		
2	1	2	2	2	1					11328	24		
7	1	1	1	1	1							3960	0
6	1	1	1	2	1							13848	24
5	1	1	1	3	1							27720	0
5	1	1	2	2	1							41544	72
4	1	1	1	4	1							34632	36
4	1	1	2	3	1							69264	72
4	1	2	2	2	1							103860	180
3	1	1	3	3	1							92400	0
3	1	2	2	3	1							138528	144
3	2	2	2	2	1							207744	312
2	2	2	2	2	2							311658	804

2.3. Vérification des résultats

Dans le but de vérifier l'exactitude des calculs nous comparons nos résultats 1°) avec ceux obtenus à partir de la méthode de Pólya[18-19] pour le système $C_n X_{m_1} Y_{m_2} Z_{m_3}$ et 2°) en comptant les graphes dessinés, pour déterminer les nombres de squelettes chiraux $A_c(n,m_1,...,m_i,...m_q)$ et de squelettes achiraux $A_a(n,m_1,...,m_i,...m_q)$ des systèmes $C_n X_{m_1} Y_{m_2} Z_{m_3}$, $C_n X_{m_1} Y_{m_2} Z_{m_3} U_{m_4}$, $C_n X_{m_1} Y_{m_2} Z_{m_3} U_{m_4} V_{m_5}$ et $C_n X_{m_1} Y_{m_2} Z_{m_3} U_{m_4} V_{m_5} W_{m_6}$, où $3 \leq n \leq 4$, $m_1+m_2+m_3+...+m_q=2n$ et $3 \leq q \leq 6$.

2.3.1. Vérification par la méthode de Pólya

Les fonctions de Pólya des dénombrements topologique $F_n^t(x,...y,...z) = \sum_{m_1,...,m_i,...,m_q} A^t(n,m_1,...,m_i,...,m_q)x^{m_1}...y^{m_i}...z^{m_q}$ et énantiomérique $F_n^e(x,...y,...,z) = \sum_{m_1,...,m_i,...,m_q} A^e(n,m_1,...,m_i,...,m_q)x^{m_1}...y^{m_i}...z^{m_q}$ des graphes des dérivés hétéropolysubstitués d'ordre 2 ou 3 du cyclopropane ($C_3 X_{m_1} Y_{m_2} Z_{m_3}$) sont données respectivement ci après :

Pour $n=3$

$F_3^t(x,...y,...z) = x^6 + x^5 y + 3x^4 y^2 + 3x^4 yz + 3x^4 z^2 + 3x^3 y^3 + 6x^3 y^2 z + 6x^3 yz^2 + 3x^3 z^3 + 3x^2 y^4 + 6x^2 y^3 z + 11x^2 y^2 z^2 + ...$

$F_3^e(x,...y,...,z) = x^6 + x^5 y + 4x^4 y^2 + 5x^4 yz + 4x^3 y^3 + 10x^3 y^2 z + 4x^3 z^3 + 4x^2 y^4 + 10x^2 y^3 z + 18x^2 y^2 z^2 + 10x^2 yz^3 + ...$

L'inventaire de Pólya des graphes des séries de composés hétéropolysubstitués d'ordre 2 ou 3 du cyclobutane ($C_4 X_{m_1}...Y_{m_i}...Z_{m_q}$), donne :

Pour $n=4$

$F_4^t(x,...y,...z) = x^8 + x^7 y + 5x^6 y^2 + 5x^5 y^3 + 13x^5 y^2 z + 10x^4 y^4 + 21x^4 y^3 z + 37x^4 y^2 z^2 + 40x^3 y^3 z^2 + ...$

$F_4^e(x,...y,...,z) = x^8 + x^7 y + 6x^6 y^2 + 7x^6 yz + 7x^5 y^3 + 21x^5 y^2 z + 13x^4 y^4 + 35x^4 y^3 z + 60x^4 y^2 z^2 + 35x^4 yz^3 + 70x^3 y^3 z^2 + ...$

En tenant compte uniquement des coefficients des termes ayant des puissances identiques en xy ou en xyz dans les expressions de $F_n^t(x,...y,...z)$ et $F_n^e(x,...y,...,z)$, on peut déduire par la méthode des différences les valeurs de $A_c(n,m_1,...,m_i,...m_q)$ et $A_a(n,m_1,...,m_i,...m_q)$ à partir des relations (2.10)-(2.11) ci-après :

$$A_c(n,m_1,...,m_i,...m_q) + A_a(n,m_1,...,m_i,...m_q) = A^t(n,m_1,...,m_i,...,m_q) \qquad (2.10)$$

48

$$2A_c(n,m_1,\ldots,m_i,\ldots m_q)+A_a(n,m_1,\ldots,m_i,\ldots m_q)= A^e(n,m_1,\ldots,m_i,\ldots,m_q) \qquad (2.11)$$

Les relations (2.10)-(2.11) lient les coefficients des 2 types de fonctions génératrices.

Exemple d'application 1 :

* Posons pour le système $C_3X_3Y_2Z$, $(n,m_1,m_2,m_3)=(3,3,2,1)$

$$\begin{cases} A_c(3,\, 3,\, 2,\, 1)+A_a(3,\, 3,\, 2,\, 1)=6 \\ 2A_c(3,\, 3,\, 2,\, 1)+A_a(3,\, 3,\, 2,\, 1)=10 \end{cases} \Longrightarrow \begin{cases} A_c(3,\, 3,\, 2,\, 1)=4 \\ A_a(3,\, 3,\, 2,\, 1)=2 \end{cases}$$

* De même pour $C_3X_2Y_2Z_2$, $(n,m_1,m_2,m_3)=(3,2,2,2)$

$$\begin{cases} A_c(3,\, 2,\, 2,\, 2)+A_a(3,\, 3,\, 2,\, 2)=11 \\ 2A_c(3,\, 2,\, 2,\, 2)+A_a(3,\, 3,\, 2,\, 2)=18 \end{cases} \Longrightarrow \begin{cases} A_c(3,\, 2,\, 2,\, 2)=7 \\ A_a(3,\, 2,\, 2,\, 2)=4 \end{cases}$$

Exemple d'application 2 :

* Posons pour le système $C_4X_5Y_2Z$, $(n,m_1,m_2,m_3)=(4,5,2,1)$

$$\begin{cases} A_c(4,\, 5,\, 2,\, 1)+A_a(4,\, 5,\, 2,\, 1)=13 \\ 2A_c(4,\, 5,\, 2,\, 1)+A_a(4,\, 5,\, 2,\, 1)=21 \end{cases} \Longrightarrow \begin{cases} A_c(4,\, 5,\, 2,\, 1)=8 \\ A_a(4,\, 5,\, 2,\, 1)=5 \end{cases}$$

* Et pour $C_4X_4Y_2Z_2$, $(n,m_1,m_2,m_3)=(4,4,2,2)$

$$\begin{cases} A_c(4,\, 4,\, 2,\, 2)+A_a(4,\, 4,\, 2,\, 2)=37 \\ 2A_c(4,\, 4,\, 2,\, 2)+A_a(4,\, 4,\, 2,\, 2)=60 \end{cases} \Longrightarrow \begin{cases} A_c(4,\, 4,\, 2,\, 2)=23 \\ A_a(4,\, 4,\, 2,\, 2)=14 \end{cases}$$

Nous notons que les valeurs de $A_c(n,m_1,m_2,m_3)$ et $A_a(n,m_1,m_2,m_3)$ sont identiques à celles du tableau 9 pour les deux exemples d'application traités ci-avant.

Les coefficients des termes $x^{m_1}\ldots y^{m_i}\ldots z^{m_q}$, $x^{m_i}\ldots y^{m_1}\ldots z^{m_q}$, $x^{m_q}\ldots y^{m_i}\ldots z^{m_1}$ et $x^{m_1}\ldots y^{m_q}\ldots z^{m_i}$ tels que $\sum_{i=1}^{q} m_i = 2n$ sont identiques en vertu de la notion de complémentarité des degrés de substitution partiels établie au chapitre 1. Par conséquent, en utilisant les données des tableaux 2.1 pour calculer $A^t(n,m_1,\ldots,m_i,\ldots,m_q)$ et $A^t(n,m_1,\ldots,m_i,\ldots,m_q)$, on voit que les valeurs obtenues par la méthode de Pólya et celles déterminées par la méthode combinatoire directe sont identiques.

2.3.2. Vérification par la méthode de comptage des graphes

La deuxième méthode de vérification de la fiabilité de nos calculs c'est de dessiner les graphes de toutes les paires d'énantiomères et des formes achirales et les compter pour obtenir respectivement $A_c(n,m_1,...,m_i,...m_q)$ et $A_a(n,m_1,...,m_i,...m_q)$. Ceci est possible pour les petites valeurs de n et m.

Nous représentons respectivement dans les figures 17 à 23 quelques exemples de graphes moléculaires des systèmes $C_n X_{m_1} Y_{m_2} Z_{m_3}$ et $C_n X_{m_1} Y_{m_2} Z_{m_3} U_{m_4}$, où $3 \leq n \leq 4$ et $m_1 + m_2 + m_3 + ... + m_q = 2n$ et $3 \leq q \leq 6$.

Pour représenter une série de graphes d'une molécules $C_n X_{m_1} ... Y_{m_i} ... Z_{m_q}$, on représente d'abord les squelettes chiraux et achiraux d'un système $C_n H_{2n-m_1} X_{m_1}$ où m_1 est le plus grand degré de substitution. Ensuite, on permute les $m_2, ..., m_q$ substituants $Y, ...,$ Z restants. Chaque graphe de $C_n H_{2n-m_1} X_{m_1}$ forme une classe à partir de laquelle on permute les autres types de substituants restants.

2.3.2.1. Représentation graphique de quelques résultats du système $C_n X_{m_1} Y_{m_2} Z_{m_3}$

Cas n=3, m$_1$=4, m$_2$=1, m$_3$=1

A_c=2

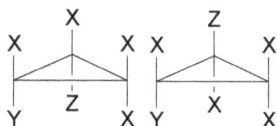

A_a=1

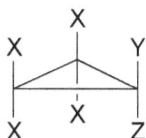

Figure 17 Graphes des squelettes chiraux et achiraux du système moléculaire C_3X_4YZ

Cas n=3, m_1=3, m_2=2, m_3=1

A_c=4

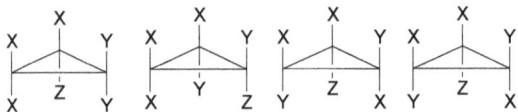

A_a=2

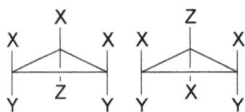

Figure 18 Graphes des squelettes chiraux et achiraux du système moléculaire $C_3X_3Y_2Z$

Cas n=3, m_1=2, m_2=2, m_3=2

A_c=7

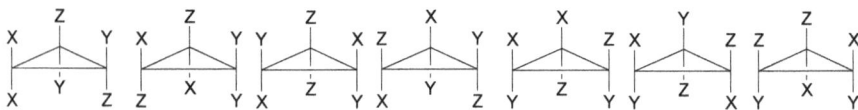

A_a=4

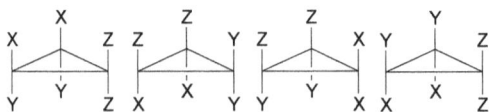

Figure 19 Graphes des squelettes chiraux et achiraux du système moléculaire *$C_3X_2Y_2Z_2$*

Cas n=4, m_1=6, m_2=1, m_3=1

A_c=2

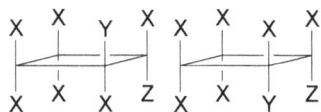

A_a=3

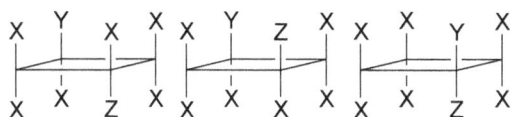

Figure 20 Graphes des squelettes chiraux et achiraux du système moléculaire C_4X_6YZ

Cas n=4, m_1=4, m_2=2, m_3=2

A_c=23

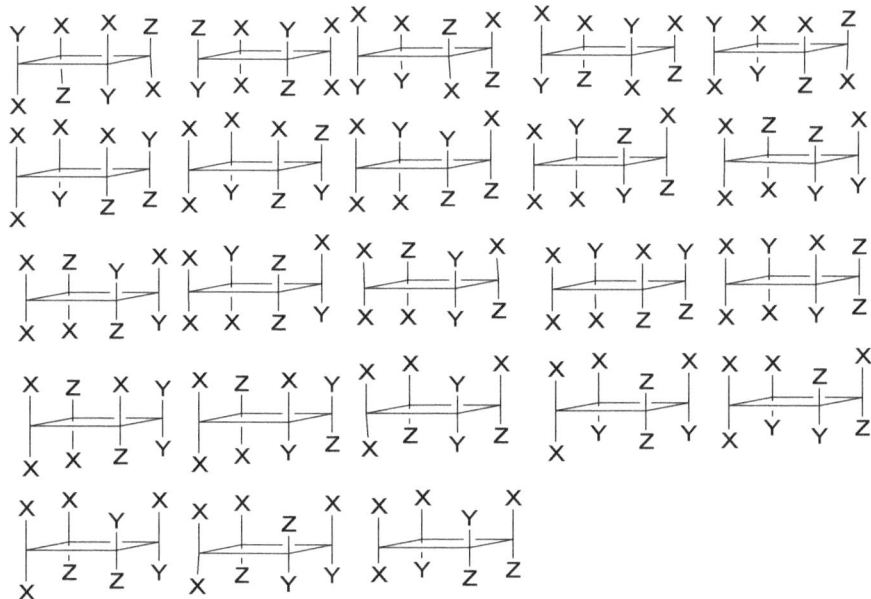

52

A_a=14

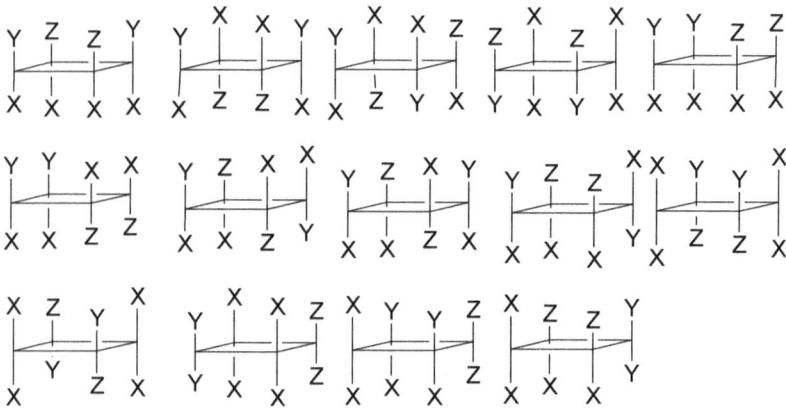

Figure 21 Graphes des squelettes chiraux et achiraux du système moléculaire $C_4X_4Y_2Z_2$

2.3.2.2. Représentation graphique de quelques résultats du système $C_n X_{m_1} Y_{m_2} Z_{m_3} U_{m_4}$

Cas n=3, m_1=3, m_2=1, m_3=1, m_4=1

A_c=10

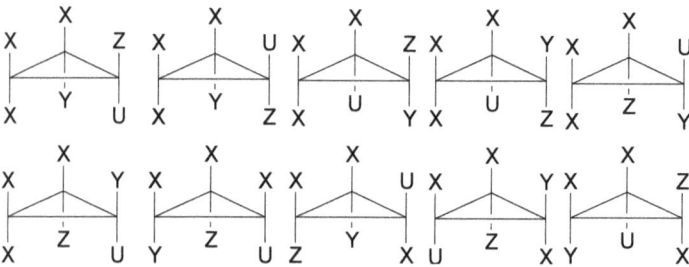

A_a=0

Figure 22 Graphes des squelettes chiraux et achiraux du système moléculaire C_3X_3YZU

Cas n=3, m_1=2, m_2=2, m_3=1, m_4=1

A_c=14

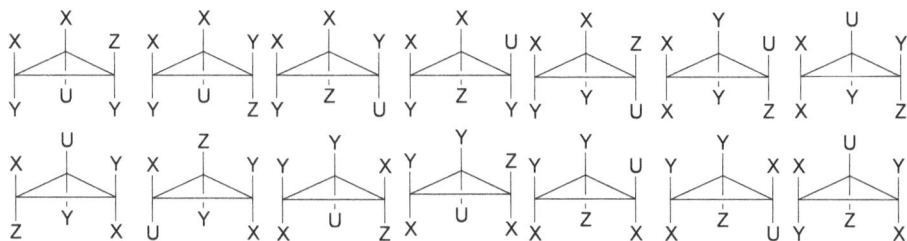

A_a=2

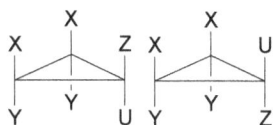

Figure 23 Graphes des squelettes chiraux et achiraux du système moléculaire $C_3X_2Y_2ZU$

Dans ce dénombrement des graphes des stéréoisomères du système moléculaire $C_n X_{m_1}...Y_{m_i}...Z_{m_q}$, nous ne considérons pas les battements de cycle qui sont engendrés par les déformations d'angles de valence et des angles de torsion.

Globalement on note une croissance exponentielle du nombre de squelettes chiraux corrélativement à l'augmentation du nombre q de type de substituants distincts composant le système moléculaire $C_n X_{m_1}...Y_{m_i}...Z_{m_q}$. Ceci s'explique par le fait qu'au fur et à mesure que q croit le nombre de plans de symétrie diminue dans les différents dérivés substitués du cycloalcane.

2.4. Exemple de génération de bibliothèques de molécules

En reprenant les graphes du système $C_n X_{m_1}...Y_{m_i}...Z_{m_q}$ pour les petits cycles de taille $n=3$ et 4 et en remplaçant les substituants X, Y, U, V, W, ou Z par des atomes ou groupes d'atomes non isomérisables tels que Cl, I, F, Br, CH_3, COOH, $COOR_1$, OH, OR_1; COCl, $CONR_1R_2$, SH, C_6H_5, ..., etc. (R_1 et R_2 sont des radicaux alkyles) nous pouvons générer des bibliothèques de molécules du cyclopropane ou du cyclobutane hétéropolysubstitué.

54

Si nous prenons par exemple le cas des systèmes $C_4X_4Y_2Z_2$ dont les graphes des stéréoisomères sont représentés dans la figure 21, en fixant $X=H$, $Y=C_6H_5$ et $Z=COOH$, nous générons la bibliothèque de molécules de la série $C_4H_4(C_6H_5)_2(COOH)_2$ présentée dans la figure 24 ci après :

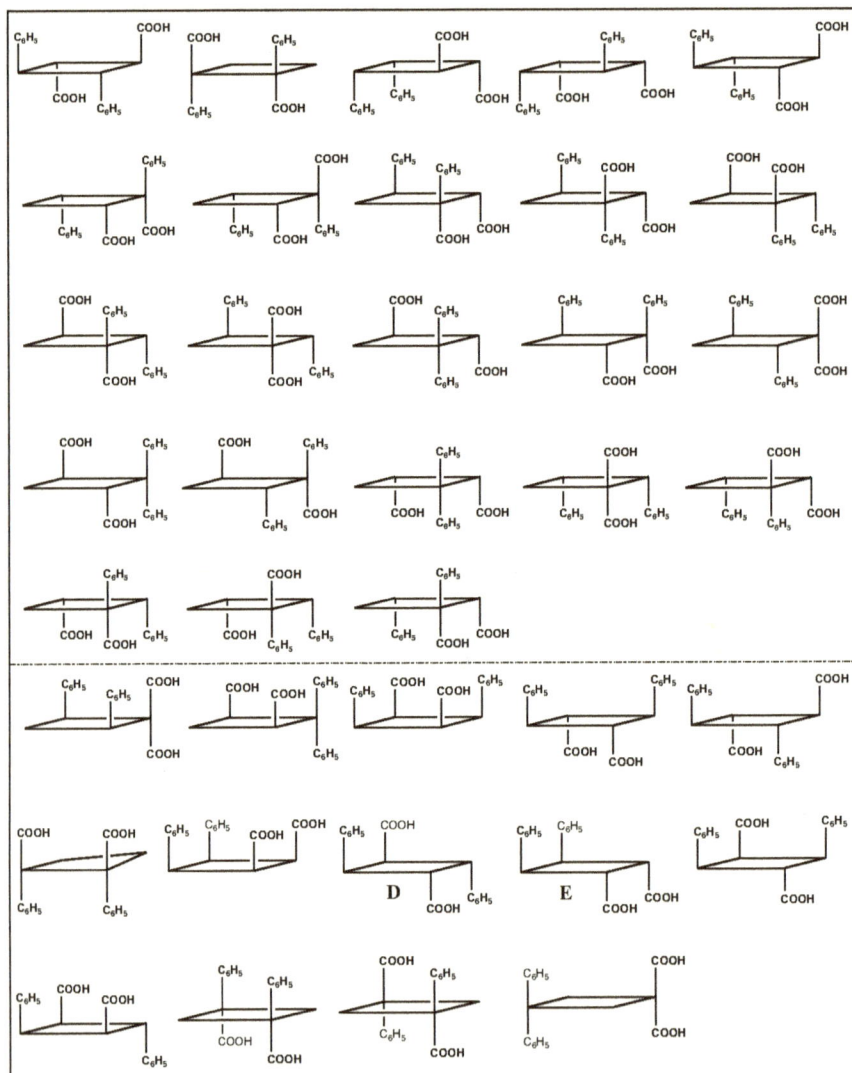

Figure 24 Bibliothèque de molécules du $C_4H_4(C_6H_5)_2(COOH)_2$

Les composés connus de la bibliothèque de molécules du $C_4H_4(C_6H_5)_2(COOH)_2$ sont l'acide α-truxilique (Fig. 24 **D**) et l'acide β-truxunique (Fig. 24 **E**). Ces deux catégories de diacides existent à l'état naturel dans les feuilles de coca ou sont obtenues par photodimérisation topochimique des formes α et β de l'acide cinnamique.[49]

De même, dans le cas du système $C_3X_2Y_2ZU$ dont les graphes des stéréoisomères sont représentés dans la figure 23, en remplaçant les substituants X par H, Y par CH_3, Z par R et U par $COOH$, nous générons la bibliothèque de molécules de la série $C_3H_2(CH_3)_2(R)(COOH)$ donnée dans la figure 25 ci après :

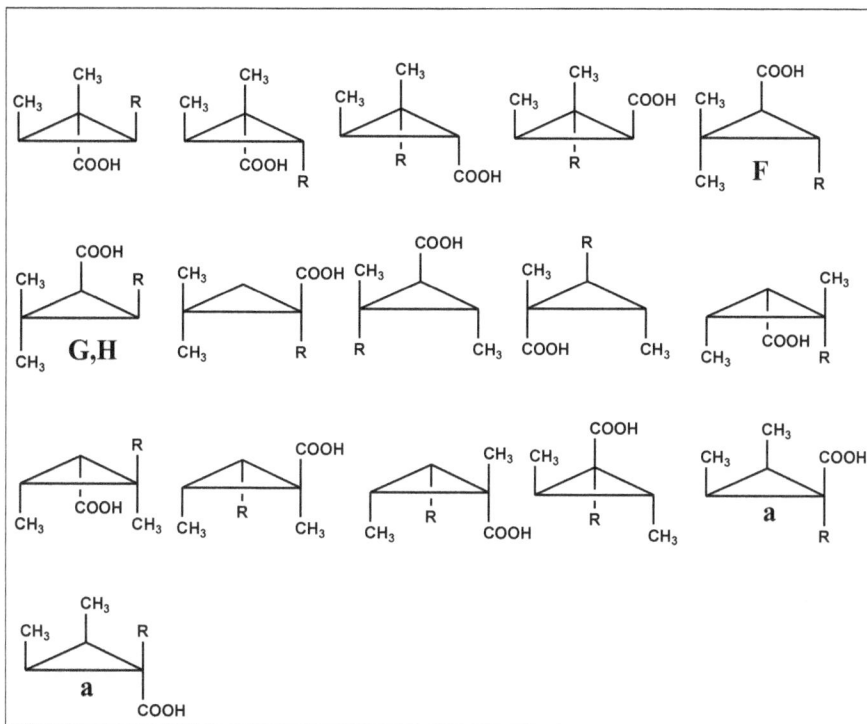

Figure 25 Bibliothèque de molécules du $C_3H_2(CH_3)_2(R)(COOH)$

La lettre (a) placée en dessous d'un graphe désigne un graphe achiral. Tous les autres graphes non identifiés sont chiraux.

* Si $R = CHO$ on a la bibliothèque de la série $C_3H_2(CH_3)_2(CHO)(COOH)$ dont les composés connus sont le trans caronaldehyde (Fig. 25 **F**) et le cis caronaldehyde (Fig. 25 **G**) qui sont utilisés pour la synthèse des dérivés des lactols ;[46]

* Si $R=CHCl_2$ on a la bibliothèque de la série $C_3H_2(CH_3)_2(CHCl_2)(COOH)$ dont le composé connu est l'acide permethrinique (fig. 25 **H)** qui entre dans la fabrication de l'insecticide permethrine.[46]

Bien que quelques molécules des bibliothèques présentées ci-haut, à titre d'exemple, sont référencées dans la littérature[54-55], il est à espérer que plusieurs d'entre-elles seront découvertes dans la nature ou mises en évidence par voie de synthèse. Par contre, d'autres ne pourront jamais exister compte tenu de leur haut degré d'instabilité du point de vue thermodynamique. En chimie combinatoire, on applique souvent la méthode *HTS* (*High Throughput Screening*) [56] c'est à dire le criblage ultra-rapide pour éliminer ces derniers de la bibliothèque ainsi que des molécules de moindre importance.

Au delà des exemples illustrés dans le présent chapitre, des bibliothèques de molécules du cycloalcane hétéropolysubstitué, $C_nX_{m_i}...Y_{m_t}..Z_{m_q}$, peuvent être constituées en variant la taille $3 \leq n \leq 288$ du cycle et en choisissant les substituants X, ..., Y, ..., Z dans une série d'atomes ou de groupements d'atomes en fonctions des types de molécules recherchés ou à prévoir.

CONCLUSION GENERALE

Le modèle mathématique de dénombrement direct des stéréoisomères des cycloalcanes monocycliques hétéropolysubstitués est bâti à partir de la correspondance entre le groupe de symétrie et le groupe de permutation.

Le calcul des combinaisons avec répétition issues des permutations des substituants distincts permet de dénombrer de manière directe et sélective les squelettes chiraux et achiraux des cycloalcanes monocycliques hétéropolysubstitués. L'avantage de cette méthode est son application à un seul type de composé substitué. Cela permet de contourner les longs calculs de la procédure classique de Pólya qui comprend les deux étapes suivantes : 1°) la détermination d'un indicateur de cycle et 2°) la recherche de la fonction polynomiale ou fonction génératrice de degré $2n$ répertoriant les inventaires des graphes de plusieurs séries de composés substitués. Les coefficients de cette fonction sont difficiles à obtenir manuellement lorsque la taille n du cycle augmente. Grâce à sa sélectivité et sa flexibilité, la méthode d'inventaire direct est économiquement avantageuse en coût calcul quelque soit la taille du cycle.

L'établissement de ce modèle mathématique de dénombrement direct pour l'inventaire des stéréoisomères des cycloalcanes monocycliques hétéropolysubstitués est une contribution pour la mise au point d'algorithme générateur de bibliothèques de structures moléculaires virtuelles ou réelles de la série des cycloalcanes monocycliques. C'est une tentative de solution à une problématique relevant de la "chimie combinatoire" qui peut nécessiter deux étapes entre autres à savoir : - la détermination d'un modèle d'inventaire adéquat qui s'établit à partir de la topologie de la molécule parente et ensuite - l'établissement des graphes des molécules à partir des combinaisons de groupements sur toutes les positions de substitutions. Cette démarche permet de faire la prévision de composés chimiques nouveaux qui seront mis en évidence par les expérimentateurs dans le processus de développement des générateurs automatiques de structures des cycloalcanes substitués à l'instar de ceux déjà utilisés en recherche moléculaire tels que le logiciel GRAAL (GénérateuR d'Arômes Alimentaires), GASP (Générateur Automatique de Structures polycycliques), … etc..[57].

REFERENCES

[1] G. Messadié, Les Grandes Découvertes de la science, *Bordas*, Paris, (1987), p. 115.

[2] A. T. Balaban, F. Harary, Early History of the Interplay between Graph Theory and Chemistry, pp. 1-4 of Chemical applications of Graph Theory, A. T. Balaban, ed., *Academic Press*, London (1976).

[3] A. Cayley, On the Analytic Forms called Trees, with Applications to the Theory of Chemical Combinations, *Rep. Brit. Soc. Adv. Sci.*, **45** (1875) 257-305 = *Math. Papers*, **9**, 427-460.

[4] A. Cayley, Ueber die Analytischen Figuren, Welche in der Mathematik Bäume Genannt Werden und Ihre Anwendun auf die Theorie Chemischer Verbindungen, *Ber. Deutsch. Chem. Ges.*, **8** (1875), 1056-1059.

[5] B. Baumslag, B. Chandler, Schuam's, Outline of Group Theory, *McGraw-Hill*, New-York, 1985.

[6] W.F. Lawvere, H. S. Schanuel, Conceptual Mathematics: A First Introduction to Categories, Cambridge University Press, Cambridge, 1997.

[7] H. R. Henze, C. M. Blair, *J. Chem. Soc.*, **53** (1931), (a) 3042-3046, (b) 3077-3085.

[8] G. Pólya, Kombinatorische Abzahlbestimmungen für Gruppen, Graphen und chemische Verbindungen, *Acta Math.* **68** (1937), 145-254. Translated as G. Pólya and R.C.Read, Combinatorial Enumeration of Groups, Graphs and Chemical Compounds, *Springer Verlag*, NY, 1987, pp.58 – 74.

[9] W. Burnside, Theory of Groups of Finite Order, *Cambridge Univ. Press*, London, 2nd Ed., (1911), p. 191.

[10] R. Otter, *Ann. Math.*, **49** (1948), 583.

[11] R. W. Robinson, F. Harary, A. T. Balaban, The number of chiral and achiral alkanes and monosubstituted alkanes, *Tetrahedron*, **32** (1976), 355-361.

[12] K. Balasubramanian, A Generalized Wreath Product method for the Enumeration of Stereo and Position Isomers of Poly-substituted Organic Compounds, *Theoretica Chimica Acta*, **51** (1979), 37-54.

[13] (a) A.T. Balaban, Chemical graphs. XXXII. Constitutional and steric Isomers of Substituted Cycloalkanes, *CROA. CHEM. ACTA*, **51** (1978), 35-42. (b) A. T. Balaban, J. W. Kennedy, L. V. Quintas, The Number of Alkanes having n Carbon and a Longest Chain of Length d, an application of a theorem of Pólya, *J. Chem. Educ.*, **65** (1988), 304-313.

[14] S. J. Cyvin, J. Brunvoll, B. N. Cyvin, E. Brendsdal, Enumeration of Isomers and

Conformers: A complete Mathematical Solution for Conjugated Polyene Hydrocarbons, *Adv. Mol. Struct. Res.*, **2** (1996), 213-245.

[15] W. Hässelbarth, E. Ruch, Classification of Rearrangement Mechanisms by Means of Double Cosets and Counting Formulas for Numbers of Classes, *Theor. Chi. Acta*, **29** (1973), 259-268.

[16] A. Kerber, Enumeration Under Finite Group Action, Basic Tools, Results and Methods, *MATCH*, **46** (2002), 151-198.

[17] (a) M. V. Almsick, H. Dolhaine, H. Hönig, Efficient Agorithm to Enumerate Isomers and Diamutamers with More Than One Type of Substituent, *J. Chem. Inf. Comput. Sci.*, **40** (2000) 956-966. (b) H. Dolhaine, H. Hönig, M. V. Almsick, Sample Applications of an Algorithm for the Calculation of the Number of Isomers with More Than One Type of Achiral Substituent, *MATCH*, **39** (1999), 21-37.

[18] R. M. Nemba, F. Ngouhouo, On The Enumeration of Chiral and Achiral Skeletons of Position Isomers of Homosubstituted Monocyclic Cycloalkanes with a Ring Size n (odd or even), *Tétrahedron*, **50** (1994), 6663-6670.

[19] R. M. Nemba, F. Ngouhouo, *New J. Chem.*, **18** (1994), 1175 – 1182.

[20] R. M. Nemba, M. Fah, On the application of sieve formula to the enumeration of the stable stereo and position isomers of deoxycyclitols, *J. Chem. Inf. Comput. Sci.*, **37**, 4 (1997), 722 – 725.

[21] R. M. Nemba, A. T. Balaban, Algorithm for the direct enumeration of Chiral and achiral skeletons of homosubstituted derivatives of a monocyclic cycloalkane with a large ring size n., *J. Chem. Inf. Comput. Sci.*, **38** (1998*), 1145-1150.*

[22] R. M. Nemba, Solution Générale du problème de dénombrement des stéréoisomères d'un cycloalcane homosubstitué, *Comptes Rendus Acad. Sci.*, Série II b, **323** (1996), 773-779.

[23] R. M. Nemba, A. T. Balaban, Enumeration of Chiral and Achiral Isomers of an n-Membered Ring with n Homomorphic Alkyl Groups, *MATCH*, **46** (2002), 235-250.

[24] S. Fujita, Systematic Enumeration of Nonrigid Isomers with Given Ligand Symmetries, *J. Chem. Inf. Comput. Sci.*, **40** (2000), 135-146.

[25] S. Fujita, Maturity of Finite Groups: An Application to combinatorial Enumeration of Isomers, *Bull. Chem. Soc. Jpn*, **71** (1998), 2071-2080.

[26] S. Fujita, Subduction of Dominant Representations: for Combinatorial Enumeration, *Theor. Chim. Acta*, **91** (1995), 315-332.

[27] S. Fujita, Pseudo-point-group approach to stereochemistry and stereoisomerism of cyclohexane derivatives, *Proc. Japan Acad.*, **77**, Ser B (2001), 197-202.

[28] S. Fujita, Subduction of Coset Representations: An Application to Enumeration of Chemical Structures, *Theor. Chim. Acta*, **76** (1989), 247-268.

[29] S. El-Basil, Prolegomenon on Theory and Applications of Tables of Marks, *MATCH*, **46** (2002), 7-23.

[30] J. H. Redfield, Catalogue of the Known Species, recent and fossil, of the family Marginellidae, *Amer. J. Conchol.*, **7** (1872), 215-269; J. H. Redfield, The Theory of Group-reduced Distributions, *Amer. J. Math.*, **49** (1927), 433-455; J. H. Redfield, Group Theory applied to Combinatory Analysis, *Match.* **41** (2000), 7-27; J. H. Redfield, Enumeration by frame Group and range Groups, *J. Graph Theory*, **8** (1984), 205-223.

[31] R. C. Read, The enumeration of Acyclic Chemical Compounds, pp. 25-61 of A. T. Balaban, ed., Chemical Applications of Graph Theory, *Academic Press*, London, (1976).

[32] N. J. de Bruijn, Pólya's Theory of Counting, Chapter 5 in Applied Combinatorial Mathematics, edited by Edwin F. Beckenbach, *John Wiley*, New York, (1964).

[33] P. W. Fowler, Isomer Counting using Point Group Symmetry, *J. Chem. Soc., Faraday Trans.*, **91** (1995), 2241-2247.

[34] F. Harary, E. Palmer, Graphical Enumeration, *Academic Press*, New York, (1973).

[35] P. Leroux, B. Miloudi, Généralisations de la Formule d'Otter, *Ann. Sci. Math. Québec*, 1, **16** (1992), 53-80.

[36] (a) United States Patents Application : 0030236248 ; (b) OMPI, Accord sur les Aspects des Droits de Propriété Intellectuelle qui touchent au Commerce (Accord ADPIC-1994), Art. 27 à 38.

[37] W. V. Metanomski, Unusual names assigned to chemical Substances, *Chemistry International*, **9** (1987), 211-216.

[38] J. D. Roberts, Chimie Organique Moderne, *Inter-ed.,* Paris, (1977), pp. 55-105.

[39] S. Beckman, H. Geiger, *Methoden Org. Chem.*, Houben-Weyl, **4** (1971), 445-478.

[40] P. Matthews, Advanced Chemistry, *Cambridge Univ. Press*, Cambridge, (1992), pp. 497-909.

[41] V. Potapov, S. Tatarintchik, Chimie Organique, *Editions MIR*, Moscou, (1981), pp. 47-77.

[42] H. –H. Otto, *Dtsch. Apoth.-Ztg.*, **115** (1975), 89-94.

[43] N. L. Biggs, Discrete Mathematics, *Oxford Univ. Press*, New York, 5th Ed., (1989), pp. 158-225.

[44] A. Kaufmann, Introduction à la combinatoire en vue des Applications, *Dunod*, Paris, 1968, pp. 196-285.

[45] R. M. Nemba, A. Emadak, Direct Enumeration of Chiral and Achiral Graphs of a Polyheterosubstituted Monocyclic Cycloalkane, *J. Int. Seq.*, **5** (2002), 1-9.

[46] R. M. Nemba, A. Emadak, Algorithme de dénombrement des graphes chiraux et achiraux d'un cycloalcane monocyclique hétérosubstitué de formule brute $C_n H_{m_1} X_{m_2} Y_{m_3}$, *C. R. Chimie Acad. Sci.*, **5** (2002), 1-7.

[47] B. W. Char, Maple 8 Learning Guide, Waterloo Maple, Inc., (2002).

[48] R. C. West, Handbook of Chemistry and Physics, CRC Press, Inc., 58th Ed., Boca Raton, FL, (1977-1978).

[49] E. L. Eliel, S. H. Wilen, L. N. Mander, in "Stereochemistry of organic compounds", *A Wiley Interscience Publication*, ed. J. Wiley and Sons Inc., New-York, (1994), p.1 – 20.

[50] Database at http://www.colby.edu/chemistry/cmp/cmp.html consulted on 12, 20th 2013.

[51] R. C. West, Handbook of Chemistry and Physics, CRC Press, Inc., 65th Ed., Boca Raton, FL, (1984-1985).

[52] J. B Pedley, Rylance, *J. Computer Analysed Thermochemical Data: Organic and Organometallic Compounds*, University of Sussex, Brighton, England, **1** (1977).

[53] The Merck Index, Chemistry constant, 30th Ed., Cambridge Soft, (2002).

[54] CS ChemDraw Ultra in ChemOffice 7.0 for Windows: Desktop to Enterprise Solutions, Cambridge Soft, (2002); www.cambridgesoft.com.

[55] Chemical data at http://webbook.nist.gov/chemistry/

[56] http://fr.wikipedia.org/wiki/Criblage_%C3%A0_haut_d%C3%A9bit, consulté le 06 avril 2014.

[57] R. Barone, Chimie Combinatoire at http://pro.chemist.online.fr/retro/barone_bis.htm, consulté le 06 avril 2014.

ANNEXES

Abréviations et acronymes

a :   Achiral.

c :   Chiral.

Cf. :   Confère

Eq.:   Equation ou Relation.

Fig.:   Figure.

rp: Rotation propre.

ri: Rotation impropre.

Définition des principaux symboles utilisés

CHAPITRE 1

C_nH_{2n} : Formule empirique du cycloalcane monocyclique parent.

$C_nH_{2n-m}X_m$: Formule brute du cycloalcane monocyclique homosubstitué par m groupements X non isomérisables.

n : Entier naturel supérieur à 2 désignant la taille du cycle ou nombre de carbones de la chaîne cyclique.

2n : Nombre total de sites de substitution du cycloalcane monocyclique parent.

G : Stéréographe ou graphe tridimensionel du cycloalcane.

D_{nh} : Groupe de symétrie ponctuel du cycloalcane monocyclique.

X : Entités chimiques ou substituants non isomérisables symbolisant les atomes ou groupes d'atomes.

m : entier indiquant le nombre de substituants non isomérisables de même type X dans le système $C_nH_{2n-m}X_m$ ou degré de substitution.

E : Opération identité de D_{nh}.

C_n^r : Opération rotation propre de D_{nh}.

$S_n^{r'}$: Opération rotation impropre de D_{nh}.

σ_h, σ_v, σ_d : Plans de symétrie de D_{nh}.

D : Diviseur de n.

i : Longueur de cycle de permutation.

j :	Nombre de cycle de permutation.
f_i^j :	Notation par indice de cycle de j permutations de longueur i.
$[i^j]$:	Notation par partition de j permutations de longueur i.
a_d, a_d' :	Coefficients des termes $[i^j]$, $[1^2 2^{n-1}]$ et $[1^4 2^{n-2}]$.
$O_{D_{nh}}(G)$:	Opérateur de symétrie de D_{nh} agissant sur G.
P :	Ensemble contenant les permutations engendrées par les 4n opérations de symétrie de D_{nh} (Les rotations propres et impropres, les plans de symétrie).
P' :	Ensemble contenant les permutations engendrées par les 2n opérations de symétrie de D_{nh} (Les rotations propres).
φ :	Fonction totient de Euler.
θ :	Entier naturel impair.
$\alpha, \beta, \gamma, \omega_i$:	Entiers naturels premiers.
$\nu, \lambda, \varepsilon, \mu, \lambda_i$:	Entiers naturels.
SS :	Ensemble des sites de substitution.
$\mathcal{D}_n$:	Ensemble des diviseurs de n.
$\mathcal{D}_{2n}$:	Ensemble des diviseurs de 2n.
$\mathcal{D}_m$:	Ensemble des diviseurs de m.
A_a :	Nombre de stéréoisomères de position achiraux.
A_c :	Nombre de stéréoisomères de position chiraux.

CHAPITRE 2

$C_n H_{m_0} X_{m_1} .. Y_{m_t} .. Z_{m_q}$:	Formule brute du cycloalcane monocyclique hétéropolysubstitué par les groupements X, Y, …, Z non isomérisables et de nature distincte.

X,...,Y,...,Z :	Entités chimiques ou substituants non isomérisables symbolisant les atomes ou groupes d'atomes.
$m_1,...,m_i,...,m_q$:	Degrés partiels de substitution ou nombres respectifs d'entités ou de substituants de types X,...,Y,...,Z.
$1 \leq i \leq q$:	Ordre de la substitution ou nombre de substituants non hydrogènes de types distincts.
$\mathcal{D}_{m_i}$ $(1 \leq i \leq q)$:	Ensemble des diviseurs de m_i.
$\mathcal{D}_c$:	Ensemble de diviseurs commun de n et des m_i ou de 2n et des m_i.
l_i :	Nombre de choix de placement des substituants X,...,Y,...,Z sur les 2 ou 4 positions invariantes de G.
m_i' :	Nombre de choix de placement des substituants X,...,Y,...,Z dans n-1 ou n-2 boîtes.
λ:	nombre de solutions compatibles entre $(l_1,...,l_i,...,l_q)$ et $(m_1',...,m_i',...,m_q')$.
a !:	Factoriel a.
T :	Coefficient multinomial.

Définition de quelques termes utilisés

Achiral : qualificatif caractérisant un stéréoisomère ou isomère superposable à son image miroir.

Acyclique : qualificatif d'un graphe qui n'a pas de cycle ou graphe non cyclique.

Boîte : synonyme de site de substitution.

Chiral : qualificatif caractérisant un stéréoisomère ou isomère non superposable à son image dans un miroir.

Chiralité : Propriété chirale ou achirale d'une molécule.

Combinaison : Assemblage, arrangement, dans un certain ordre ou non, des éléments ou objets semblables ou différentes.

Combinaison homomorphe : Assemblage, arrangement, dans un certain ordre ou non, des éléments ou objets semblables.

Combinaison hétéromorphe : Assemblage, arrangement, dans un certain ordre ou non, des éléments ou objets dont deux ou plusieurs sont deux à deux distincts.

Cycle : synonyme de chaîne cyclique.

Cycle de permutation : uples de points classés dans un ordre déterminé par une opération de permutation.

Degré de substitution : nombre total d'atomes d'hydrogène du cycloalcane monocyclique remplacés par des atomes, groupes d'atomes ou radicaux alkyles.

Dénombrement : recensement des stéréoisomères ou isomères.

Divisibilité : qualité de ce qui est divisible sans reste.

Enantiomère : composé chimique inverse optique de son homologue.

Fusion : réunion ou combinaison additive.

Groupements : synonyme de substituants ou de radicaux alkyles.

Hétéromorphe : forme distincte ou différente.

Hétéropolysubstitution : substitution de plusieurs atomes d'hydrogène du cycloalcane monocyclique par plusieurs atomes ou groupes d'atomes non isomérisables et distincts.

Hétéropolysubstitué : qualificatif d'un cycloalcane monocyclique contenant des atomes ou groupements d'atomes non isomérisables et distincts.

Homomorphe : forme identique ou semblable.

Homopolysubstitution : substitution de plusieurs atomes d'hydrogène du cycloalcane monocyclique par plusieurs atomes ou groupes d'atomes non isomérisables et semblables.

Homopolysubstitué : qualificatif d'un cycloalcane monocyclique contenant des atomes ou groupes d'atomes non isomérisables et identiques.

Inventaire : synonyme de dénombrement.

Longueur du cycle de permutation : nombre d'éléments composant une permutation.

Monocycle : synonyme de chaîne carbonée du cycloalcane monocyclique.

Ordre d'une substitution : nombre de substituants de types distincts non hydrogènes.

Permutation : passage d'un ordre d'arrangement ou de combinaison de *m* éléments à un autre ordre d'arrangement ou de combinaison distinct des mêmes éléments.

Site de substitution : point ou carbone portant 2 atomes d'hydrogène tel que illustré dans le stéréographe du cycloalcane monocyclique. Au sens figuré, ce carbone est assimilé à une boîte contenant deux objets (atomes d'hydrogène) pouvant être substitués par d'autres objets distincts.

Squelette : synonyme de stéréoisomères.

Stéréographe : graphe tridimensionnel.

Stéréoisomérie : isomérie stérique.

Stéréoisomère : dénomination d'un composé chimique en isomère stérique.

Substituant : entité chimique utilisée pour remplacer ou encore substituer un atome d'hydrogène du cycloalcane monocyclique.

Substitution : réaction chimique dans laquelle un atome A d'une molécule est remplacé par un autre atome ou groupe d'atomes B.

Taille du cycle : nombre de carbone sur la chaîne cyclique d'un cycloalcane.

Programmes de calcul en langage Maple

1) PROGRAMME DE CALCUL EN LANGAGE MAPLE DU NOMBRE DES GRAPHES DES STEREOISOMERES DU SYSTEME $C_n H_{2n-m} X_m$

2) PROGRAMME DE CALCUL EN LANGAGE MAPLE DU NOMBRE DE GRAPHES DES STEREOISOMERES DU CYCLOALCANE DE FORMULE BRUTE $C_n X_{m_1} Y_{m_2} Z_{m_3}$ AYANT UNE HETEROPOLYSUBSTITUTION BINAIRE OU TERNAIRE

3) PROGRAMME DE CALCUL EN LANGAGE MAPLE DU NOMBRE DE GRAPHES DES STEREOISOMERES DU CYCLOALCANE DE FORMULE BRUTE $C_n X_{m_1} Y_{m_2} Z_{m_3} U_{m_4}$ AYANT UNE HETEROPOLYSUBSTITUTION TERNAIRE OU QUATERNAIRE

4) PROGRAMME DE CALCUL EN LANGAGE MAPLE DU NOMBRE DE GRAPHES DES STEREOISOMERES DU CYCLOALCANE DE FORMULE BRUTE $C_n X_{m_1} Y_{m_2} Z_{m_3} U_{m_4} V_{m_5}$ AYANT UNE HETEROPOLYSUBSTITUTION QUATERNAIRE OU QUINTUPLET

5) PROGRAMME DE CALCUL EN LANGAGE MAPLE DU NOMBRE DE GRAPHES DES STEREOISOMERES DU CYCLOALCANE DE FORMULE BRUTE $C_n X_{m_1} Y_{m_2} Z_{m_3} U_{m_4} V_{m_5} W_{m_6}$ AYANT UNE HETEROPOLYSUBSTITUTION QUINTUPLET OU SEXTUPLET

MAPLE PROGRAM FOR THE COMPUTATION OF THE NUMBERS OF CHIRAL AND ACHIRAL SKELETONS OF HOMOPOLYSUBSTITUTED MONOCYCLIC CYCLOALKANES CnH2n-mXm

n = ring size, m = number of non isomerizable substituents

By Alphonse EMADAK

> restart:

Subprogram: To compute of the numbers of permutations induced by proper and improper symmetries of Dnh

> Ad1:=proc(x1)

> local y1;

> if type(x1,odd) or x1=2 then y1:=numtheory[phi](x1)

>   else   if   type(x1/2,odd)   and   x1<>2   then y1:=2*numtheory[phi](x1)+numtheory[phi](x1/2);

>   else y1:=2*numtheory[phi](x1)

>   fi;

> fi;

> end:

Subprogram: To compute of the numbers of permutations induced by proper symmetries of Dnh

> Ad2:=proc(x2)

> local y2;

> if type(x2,odd) or type(x2,even) then y2:=numtheory[phi](x2);

> fi;

> end:

Subprogram: To compute the number Ac and Aa respectively of chiral and achiral graphs of homopolysubstituted skeletons of cycloalkanes

> stereo:=proc(n,m) local Dn,Dm,Dc,S1,S2: # CASE n ODD and m ODD

>   if   type(n,odd)   and   type(m,   odd)   then Dn:=numtheory[divisors](2*n): Dm:=numtheory[divisors](m):

> Dc:= convert((Dn intersect Dm) minus {2}, list):

>   S1:=sum('(2*Ad2(Dc[i])-Ad1(Dc[i]))*binomial(2*n/Dc[i],m/Dc[i])','i'=1..nops(Dc)):

```
>                                              S2:=sum('(Ad1(Dc[i])-
Ad2(Dc[i]))*binomial(2*n/Dc[i],m/Dc[i])','i'=1..nops(Dc)):

>Ac(n,m):=(1/(4*n))*(S1-2*n*binomial(n-1,(m-1)/2));

>Aa(n,m):=(1/(2*n))*(S2+2*n*binomial(n-1,(m-1)/2));

>print (Ac(n,m),Aa(n,m));

>fi;                              # CASE n ODD and m EVEN

>     if      type(n,odd)      and      type(m,      even)      then
Dn:=numtheory[divisors](2*n): Dm:=numtheory[divisors](m):

>Dc:= convert((Dn intersect Dm) minus {2}, list):

>                                              S1:=sum('(2*Ad2(Dc[i])-
Ad1(Dc[i]))*binomial(2*n/Dc[i],m/Dc[i])','i'=1..nops(Dc)):

>                                              S2:=sum('(Ad1(Dc[i])-
Ad2(Dc[i]))*binomial(2*n/Dc[i],m/Dc[i])','i'=1..nops(Dc)):

>Ac(n,m):=(1/(4*n))*((S1-binomial(n,m/2)));

>Aa(n,m):=(1/(2*n))*(S2+(n+1)*binomial(n,m/2));

>print (Ac(n,m),Aa(n,m));

>fi;                              # CASE n EVEN and m ODD

>if type(n,even) and type(m, odd) then Dn:=numtheory[divisors](n):
Dm:=numtheory[divisors](m):

>Dc:= convert((Dn intersect Dm) minus {2}, list):

>                                              S1:=sum('(2*Ad2(Dc[i])-
Ad1(Dc[i]))*binomial(2*n/Dc[i],m/Dc[i])','i'=1..nops(Dc)):

>                                              S2:=sum('(Ad1(Dc[i])-
Ad2(Dc[i]))*binomial(2*n/Dc[i],m/Dc[i])','i'=1..nops(Dc)):

>Ac(n,m):=(1/(4*n))*(S1-2*n*binomial(n-1,(m-1)/2));

>Aa(n,m):=(1/(2*n))*(S2+2*n*binomial(n-1,(m-1)/2));

>print (Ac(n,m),Aa(n,m));

>fi;                              # CASE n EVEN and m EVEN

>     if      type(n,even)      and      type(m,      even)      then
Dn:=numtheory[divisors](n): Dm:=numtheory[divisors](m):

>Dc:= convert((Dn intersect Dm) minus {2}, list):

>                                              S1:=sum('(2*Ad2(Dc[i])-
Ad1(Dc[i]))*binomial(2*n/Dc[i],m/Dc[i])','i'=1..nops(Dc)):

>                                              S2:=sum('(Ad1(Dc[i])-
Ad2(Dc[i]))*binomial(2*n/Dc[i],m/Dc[i])','i'=1..nops(Dc)):

>                      Ac(n,m):=(1/(4*n))*(S1+(1/(n-1))*((m^2/2)-
n*(m+1)+1)*binomial(n,m/2));
```

71

```
>               Aa(n,m):=(1/(2*n))*(S2+(1/(n-1))*(n^2+n*(m+1)-(m^2/2)-
2)*binomial(n,m/2));
> print(Ac(n,m),Aa(n,m));
> fi;
> end:
```

TO RUN THE PROGRAM: Type stereo(n,m); and press Enter to get the number of chiral and achiral skeletons of CnH2n-mXm respectively.

Exemple:

```
> stereo(6,5);
```
$$28, 10$$

```
> stereo(288,56);
```
$$2951225272402716823760104518023248870347397571826207172942276441013923406322 , 364557291940453056883639518414730079392$$

```
> stereo(4,5);
```
$$2, 3$$

```
> stereo(3,3);
```
$$1, 2$$

```
> stereo(4,6);
```
$$1, 4$$

```
> stereo(5,1);
```
$$0, 1$$

```
> stereo(5,2);
```
$$2, 3$$

```
> stereo(5,3);
```
$$4, 4$$

```
> stereo(5,4);
```
$$10, 6$$

```
> stereo(5,5);
```

72

$$10, 6$$

```
> stereo(6,1);
```

$$0, 1$$

```
> stereo(6,2);
```

$$35, 20$$

```
> stereo(288,56);
```

$$2951225272402716823760104518023248870347397571826207172942276441 01 \ 3923406322, 3645572919404530568836395184147300 79392$$

```
> stereo(179,36);
```

$$52903646491153673718549147774423094080603417135,\\1154689618409470388366550$$

```
> stereo(100,50);
```

$$11346459448081151526685799035640604680857259 14,\\16961650742864790160 7828$$

```
> stereo(134,200);
```

$$8749529261666334661889447357663760539260470553803283 2051582840,\\5050450591080176093468200110 9960$$

MAPLE PROGRAM FOR THE COMPUTATION OF THE NUMBERS OF CHIRAL AND ACHIRAL SKELETONS OF HETEROPOLYSUBSTITUTED MONOCYCLIC CYCLOALKANES CnXm1Ym2Zm3

n = ring size, mi = number of non isomerizable substituents, i=1,2,3.

By Alphonse EMADAK

```
> restart:
```

Subprogram: To compute of the numbers of permutations induced by proper and improper symmetries of Dnh

```
> Ad1:=proc(x1)

> local y1;

> if type(x1,odd) or x1=2 then y1:=numtheory[phi](x1)

>               else    if    type(x1/2,odd)    and    x1<>2    then
y1:=2*numtheory[phi](x1)+numtheory[phi](x1/2);

>            else y1:=2*numtheory[phi](x1)

>         fi;

> fi;

> end:
```

Subprogram: To compute of the numbers of permutations induced by proper symmetries of Dnh

```
> Ad2:=proc(x2)

> local y2;

> if type(x2,odd) or type(x2,even) then y2:=numtheory[phi](x2);

> fi;

> end:
```

Subprogram: To compute the number Ac and Aa respectively of chiral and achiral graphs of heteropolysubstituted skeletons of cycloalkanes

```
>                 stereo:=proc(n,m1,m2,m3)                     local
Dn,Dm1,Dm2,Dm3,Dc,S1,S2,P,KT1,KT2,K11,K21,K31,K41,K51,K61,K12,
K22,K32,K42,K52,K62,K72,K82,K92,K102,K112,K122,K132,K142,K152:
# CASE n ODD

>     if      type(n,odd)      then      Dn:=numtheory[divisors](2*n);
Dm1:=numtheory[divisors](m1):          Dm2:=numtheory[divisors](m2):
Dm3:=numtheory[divisors](m3):

> Dc:= convert((Dn intersect Dm1 intersect Dm2 intersect Dm3) minus
{2}, list):

>                                         S1:=sum('(2*Ad2(Dc[i])-
Ad1(Dc[i]))*combinat[multinomial](2*n/Dc[i],m1/Dc[i],m2/Dc[i],m3/D
c[i])','i'=1..nops(Dc)):
```

```
>                                                   S2:=sum('(Ad1(Dc[i])-
Ad2(Dc[i]))*combinat[multinomial](2*n/Dc[i],m1/Dc[i],m2/Dc[i],m3/D
c[i])','i'=1..nops(Dc)):

>    if (m1-2)>=0 and type((m1-2)/2,integer) and type(m2/2,integer)
and                     type(m3/2,integer)                      then
K11:=combinat[multinomial](2,2,0,0)*combinat[multinomial](n-1,(m1-
2)/2,m2/2,m3/2): else K11:=0:fi:

>    if (m2-2)>=0 and type(m1/2,integer) and type((m2-2)/2,integer)
and                     type(m3/2,integer)                      then
K21:=combinat[multinomial](2,0,2,0)*combinat[multinomial](n-
1,m1/2,(m2-2)/2,m3/2): else K21:=0:fi:

>    if (m3-2)>=0 and  type(m1/2,integer) and type(m2/2,integer)
and                     type((m3-2)/2,integer)                  then
K31:=combinat[multinomial](2,0,0,2)*combinat[multinomial](n-
1,m1/2,m2/2,(m3-2)/2): else K31:=0:fi:

>     if (m1-1)>=0 and (m2-1)>=0 and type((m1-1)/2,integer) and
type((m2-1)/2,integer)      and      type(m3/2,integer)      then
K41:=combinat[multinomial](2,1,1,0)*combinat[multinomial](n-1,(m1-
1)/2,(m2-1)/2,m3/2): else K41:=0:fi:

>     if (m1-1)>=0 and (m3-1)>=0 and type((m1-1)/2,integer) and
type(m2/2,integer)      and      type((m3-1)/2,integer)      then
K51:=combinat[multinomial](2,1,0,1)*combinat[multinomial](n-1,(m1-
1)/2,m2/2,(m3-1)/2): else K51:=0:fi:

>     if (m2-1)>=0 and (m3-1)>=0 and type(m1/2,integer) and
type((m2-1)/2,integer)      and      type((m3-1)/2,integer)      then
K61:=combinat[multinomial](2,0,1,1)*combinat[multinomial](n-
1,m1/2,(m2-1)/2,(m3-1)/2): else K61:=0:fi:

> KT1:=K11+K21+K31+K41+K51+K61:

>            if   type(m1/2,integer)   and   type(m2/2,integer)   and
type(m3/2,integer)                                              then
P:=combinat[multinomial](n,m1/2,m2/2,m3/2): else P:=0: fi:

> Ac(n,m1,m2,m3):=(1/(4*n))*(S1+(n-1)*P)-(1/4)*KT1:

> Aa(n,m1,m2,m3):=(1/(2*n))*(S2+P)+(1/2)*KT1:

> print (Ac(n,m1,m2,m3),Aa(n,m1,m2,m3));

>                                                              else
# CASE n EVEN

>    if      type(n,even)     then      Dn:=numtheory[divisors](n);
Dm1:=numtheory[divisors](m1):      Dm2:=numtheory[divisors](m2):
Dm3:=numtheory[divisors](m3):

> Dc:= convert((Dn intersect Dm1 intersect Dm2 intersect Dm3) minus
{2}, list):

>                                                   S1:=sum('(2*Ad2(Dc[i])-
Ad1(Dc[i]))*combinat[multinomial](2*n/Dc[i],m1/Dc[i],m2/Dc[i],m3/D
```

```
c[i])','i'=1..nops(Dc)):
>                                              S2:=sum('(Ad1(Dc[i])-
Ad2(Dc[i]))*combinat[multinomial](2*n/Dc[i],m1/Dc[i],m2/Dc[i],m3/D
c[i])','i'=1..nops(Dc)):
>              if   (m1-4)>=0   and   type((m1-4)/2,integer)   and
type(m2/2,integer)          and          type(m3/2,integer)        then
K12:=combinat[multinomial](4,4,0,0)*combinat[multinomial](n-2,(m1-
4)/2,m2/2,m3/2); else K12:=0:fi:

>    if (m2-4)>=0 and type(m1/2,integer) and type((m2-4)/2,integer)
and                      type(m3/2,integer)                       then
K22:=combinat[multinomial](4,0,4,0)*combinat[multinomial](n-
2,m1/2,(m2-4)/2,m3/2); else K22:=0:fi:

>    if (m3-4)>=0 and type(m1/2,integer) and type(m2/2,integer) and
type((m3-4)/2,integer)                                            then
K32:=combinat[multinomial](4,0,0,4)*combinat[multinomial](n-
2,m1/2,m2/2,(m3-4)/2); else K32:=0:fi:

>      if  (m1-3)>=0  and  (m2-1)>=0  and  type((m1-3)/2,integer)  and
type((m2-1)/2,integer)        and        type(m3/2,integer)       then
K42:=combinat[multinomial](4,3,1,0)*combinat[multinomial](n-2,(m1-
3)/2,(m2-1)/2,m3/2); else K42:=0:fi:

>          if  (m2-3)>=0  and  (m3-1)>=0  and  type(m1/2,integer)  and
type((m2-3)/2,integer)       and       type((m3-1)/2,integer)     then
K52:=combinat[multinomial](4,0,3,1)*combinat[multinomial](n-
2,m1/2,(m2-3)/2,(m3-1)/2); else K52:=0:fi:

>      if  (m1-3)>=0  and  (m3-1)>=0  and  type((m1-3)/2,integer)  and
type(m2/2,integer)          and          type((m3-1)/2,integer)   then
K62:=combinat[multinomial](4,3,0,1)*combinat[multinomial](n-2,(m1-
3)/2,m2/2,(m3-1)/2); else K62:=0:fi:

>      if  (m1-1)>=0  and  (m2-3)>=0  and  type((m1-1)/2,integer)  and
type((m2-3)/2,integer)          and          type(m3/2,integer)   then
K72:=combinat[multinomial](4,1,3,0)*combinat[multinomial](n-2,(m1-
1)/2,(m2-3)/2,m3/2); else K72:=0:fi:

>          if  (m2-1)>=0  and  (m3-3)>=0  and  type(m1/2,integer)  and
type((m2-1)/2,integer)       and       type((m3-3)/2,integer)     then
K82:=combinat[multinomial](4,0,1,3)*combinat[multinomial](n-
2,m1/2,(m2-1)/2,(m3-3)/2); else K82:=0:fi:

>      if  (m1-1)>=0  and  (m3-3)>=0  and  type((m1-1)/2,integer)  and
type(m2/2,integer)          and          type((m3-3)/2,integer)   then
K92:=combinat[multinomial](4,1,0,3)*combinat[multinomial](n-2,(m1-
1)/2,m2/2,(m3-3)/2); else K92:=0:fi:

>      if  (m1-2)>=0  and  (m2-2)>=0  and  type((m1-2)/2,integer)  and
type((m2-2)/2,integer)          and          type(m3/2,integer)   then
K102:=combinat[multinomial](4,2,2,0)*combinat[multinomial](n-
2,(m1-2)/2,(m2-2)/2,m3/2); else K102:=0:fi:
```

```
>        if  (m2-2)>=0  and  (m3-2)>=0  and  type(m1/2,integer)  and
type((m2-2)/2,integer)       and       type((m3-2)/2,integer)    then
K112:=combinat[multinomial](4,0,2,2)*combinat[multinomial](n-
2,m1/2,(m2-2)/2,(m3-2)/2); else K112:=0:fi:

>        if  (m1-2)>=0  and  (m3-2)>=0  and  type((m1-2)/2,integer)  and
type(m2/2,integer)       and       type((m3-2)/2,integer)    then
K122:=combinat[multinomial](4,2,0,2)*combinat[multinomial](n-
2,(m1-2)/2,m2/2,(m3-2)/2); else K122:=0:fi:

>        if  (m1-1)>=0  and  (m2-1)>=0  and  type((m1-1)/2,integer)  and
type((m2-1)/2,integer)       and       type((m3-2)/2,integer)    then
K132:=combinat[multinomial](4,1,1,2)*combinat[multinomial](n-
2,(m1-1)/2,(m2-1)/2,(m3-2)/2); else K132:=0:fi:

>        if  (m1-2)>=0  and  (m2-1)>=0  and  type((m1-2)/2,integer)  and
type((m2-1)/2,integer)       and       type((m3-1)/2,integer)    then
K142:=combinat[multinomial](4,2,1,1)*combinat[multinomial](n-
2,(m1-2)/2,(m2-1)/2,(m3-1)/2); else K142:=0:fi:

>        if  (m1-1)>=0  and  (m2-2)>=0  and  type((m1-1)/2,integer)  and
type((m2-2)/2,integer)       and       type((m3-1)/2,integer)    then
K152:=combinat[multinomial](4,1,2,1)*combinat[multinomial](n-
2,(m1-1)/2,(m2-2)/2,(m3-1)/2); else K152:=0:fi:

>
KT2:=K12+K22+K32+K42+K52+K62+K72+K82+K92+K102+K112+K122+K132+K142+
K152:

>            if  type(m1/2,integer)  and  type(m2/2,integer)  and
type(m3/2,integer)                                              then
P:=combinat[multinomial](n,m1/2,m2/2,m3/2); else P:=0: fi:

>Ac(n,m1,m2,m3):=(1/(4*n))*(S1+((n/2)-1)*P)-(1/8)*KT2:

>Aa(n,m1,m2,m3):=(1/(2*n))*(S2+((n/2)+2)*P)+(1/4)*KT2:

>print (Ac(n,m1,m2,m3),Aa(n,m1,m2,m3));

>fi;

>end:

>end:
```

TO RUN THE PROGRAM: Type stereo(n,m1,m2,m3); and press Enter to get the number of chiral and achiral skeletons of CnXm1Ym2Zm3 respectively.

Exemple:

```
>stereo(3,2,2,2);
```
$$7,4$$

```
>stereo(8,3,2,11);
```
$$1338, 54$$

MAPLE PROGRAM FOR THE COMPUTATION OF THE NUMBERS OF CHIRAL AND ACHIRAL SKELETONS OF HETEROPOLYSUBSTITUTED MONOCYCLIC CYCLOALKANES CnXm1Ym2Zm3Um4

n = ring size, mi = number of non isomerizable substituents, i=1,2,3,4.

By Alphonse EMADAK

```
>
> restart:
```

Subprogram: To compute of the numbers of permutations induced by proper and improper symmetries of Dnh

```
> Ad1:=proc(x1)
> local y1;
> if type(x1,odd) or x1=2 then y1:=numtheory[phi](x1)
>              else if    type(x1/2,odd)    and    x1<>2    then
y1:=2*numtheory[phi](x1)+numtheory[phi](x1/2);
>          else y1:=2*numtheory[phi](x1)
>      fi;
> fi;
> end:
```

Subprogram: To compute of the numbers of permutations induced by proper symmetries of Dnh

```
> Ad2:=proc(x2)
> local y2;
> if type(x2,odd) or type(x2,even) then y2:=numtheory[phi](x2);
> fi;
> end:
```

Subprogram: To compute the number Ac and Aa respectively of chiral and achiral graphs of heteropolysubstituted skeletons of cycloalkanes

```
>              stereo:=proc(n,m1,m2,m3,m4)              local
Dn,Dm1,Dm2,Dm3,Dm4,Dc,S1,S2,P,KT1,KT2,K11,K21,K31,K41,K51,K61,K71,
K81,K91,K101,K12,
K22,K32,K42,K52,K62,K72,K82,K92,K102,K112,K122,K132,K142,K152,K162
,K172,K182,K192,K202,K212,K222,K232,K242,K252,K262,K272,K282,K292,
K302,K2322,K2922,K822,K2622:
# CASE n ODD

>    if     type(n,odd)     then     Dn:=numtheory[divisors](2*n);
Dm1:=numtheory[divisors](m1):         Dm2:=numtheory[divisors](m2):
Dm3:=numtheory[divisors](m3):Dm4:=numtheory[divisors](m4):
```

```
> Dc:= convert((Dn intersect Dm1 intersect Dm2 intersect Dm3
intersect Dm4) minus {2}, list):

>                               S1:=sum('(2*Ad2(Dc[i])-
Ad1(Dc[i]))*combinat[multinomial](2*n/Dc[i],m1/Dc[i],m2/Dc[i],m3/D
c[i],m4/Dc[i])','i'=1..nops(Dc)):

>                               S2:=sum('(Ad1(Dc[i])-
Ad2(Dc[i]))*combinat[multinomial](2*n/Dc[i],m1/Dc[i],m2/Dc[i],m3/D
c[i],m4/Dc[i])','i'=1..nops(Dc)):

>   if (m1-2)>=0 and type((m1-2)/2,integer) and type(m2/2,integer)
and      type(m3/2,integer)      and      type(m4/2,integer)      then
K11:=combinat[multinomial](2,2,0,0,0)*combinat[multinomial](n-
1,(m1-2)/2,m2/2,m3/2,m4/2): else K11:=0:fi:

>   if (m2-2)>=0 and type(m1/2,integer) and type((m2-2)/2,integer)
and      type(m3/2,integer)      and      type(m4/2,integer)      then
K21:=combinat[multinomial](2,0,2,0,0)*combinat[multinomial](n-
1,m1/2,(m2-2)/2,m3/2,m4/2): else K21:=0:fi:

>    if (m3-2)>=0 and   type(m1/2,integer) and type(m2/2,integer)
and      type((m3-2)/2,integer)      and      type(m4/2,integer)     then
K31:=combinat[multinomial](2,0,0,2,0)*combinat[multinomial](n-
1,m1/2,m2/2,(m3-2)/2, m4/2): else K31:=0:fi:

>    if (m4-2)>=0 and   type(m1/2,integer) and type(m2/2,integer)
and      type(m3/2,integer)      and      type((m4-2)/2,integer)     then
K41:=combinat[multinomial](2,0,0,0,2)*combinat[multinomial](n-
1,m1/2,m2/2,(m3-2)/2, m4/2): else K41:=0:fi:

>     if (m1-1)>=0 and (m2-1)>=0 and type((m1-1)/2,integer) and
type((m2-1)/2,integer)     and        type(m3/2,integer)       and
type(m4/2,integer)                                             then
K51:=combinat[multinomial](2,1,1,0,0)*combinat[multinomial](n-
1,(m1-1)/2,(m2-1)/2,m3/2,m4/2): else K51:=0:fi:

>     if (m1-1)>=0 and (m3-1)>=0 and type((m1-1)/2,integer) and
type(m2/2,integer)        and        type((m3-1)/2,integer)     and
type(m4/2,integer)                                             then
K61:=combinat[multinomial](2,1,0,1,0)*combinat[multinomial](n-
1,(m1-1)/2,m2/2,(m3-1)/2,m4/2): else K61:=0:fi:

>       if (m2-1)>=0 and (m3-1)>=0 and type(m1/2,integer) and
type((m2-1)/2,integer)      and        type((m3-1)/2,integer)   and
type(m4/2,integer)                                             then
K71:=combinat[multinomial](2,0,1,1,0)*combinat[multinomial](n-
1,m1/2,(m2-1)/2,(m3-1)/2,m4/2): else K71:=0:fi:

>       if (m1-1)>=0 and (m4-1)>=0 and type((m1-1)/2,integer) and
type(m2/2,integer)         and        type((m4-1)/2,integer)    and
type(m3/2,integer)                                             then
K81:=combinat[multinomial](2,1,0,0,1)*combinat[multinomial](n-
1,(m1-1)/2,m2/2,m3/2,(m4-1)/2): else K81:=0:fi:

> if (m2-1)>=0 and (m4-1)>=0 and type(m1/2,integer) and type((m2-
```

```
1)/2,integer)   and   type((m4-1)/2,integer)   and   type(m3/2,integer)
then
K91:=combinat[multinomial](2,0,1,0,1)*combinat[multinomial](n-
1,m1/2,(m2-1)/2,m3/2,(m4-1)/2): else K91:=0:fi:

> if  (m3-1)>=0 and  (m4-1)>=0 and type(m1/2,integer)  and type((m3-
1)/2,integer)   and   type((m4-1)/2,integer)   and   type(m2/2,integer)
then
K101:=combinat[multinomial](2,0,1,0,1)*combinat[multinomial](n-
1,m1/2,(m3-1)/2,m2/2,(m4-1)/2): else K101:=0:fi:

> KT1:=K11+K21+K31+K41+K51+K61+K71+K81+K91+K101:

>             if   type(m1/2,integer)   and   type(m2/2,integer)   and
type(m3/2,integer)         and         type(m4/2,integer)         then
P:=combinat[multinomial](n,m1/2,m2/2,m3/2,m4/2): else P:=0: fi:

> Ac(n,m1,m2,m3,m4):=(1/(4*n))*(S1+(n-1)*P)-(1/4)*KT1:

> Aa(n,m1,m2,m3,m4):=(1/(2*n))*(S2+P)+(1/2)*KT1:

> print (Ac(n,m1,m2,m3,m4),Aa(n,m1,m2,m3,m4));

>                                                             else
# CASE n EVEN

>     if     type(n,even)     then     Dn:=numtheory[divisors](n);
Dm1:=numtheory[divisors](m1):         Dm2:=numtheory[divisors](m2):
Dm3:=numtheory[divisors](m3):Dm4:=numtheory[divisors](m4):

> Dc:=  convert((Dn  intersect  Dm1  intersect  Dm2  intersect  Dm3
intersect Dm4) minus {2}, list):

>                                        S1:=sum('(2*Ad2(Dc[i])-
Ad1(Dc[i]))*combinat[multinomial](2*n/Dc[i],m1/Dc[i],m2/Dc[i],m3/D
c[i],m4/Dc[i])','i'=1..nops(Dc)):

>                                        S2:=sum('(Ad1(Dc[i])-
Ad2(Dc[i]))*combinat[multinomial](2*n/Dc[i],m1/Dc[i],m2/Dc[i],m3/D
c[i],m4/Dc[i])','i'=1..nops(Dc)):

>               if    (m1-4)>=0   and   type((m1-4)/2,integer)   and
type(m2/2,integer)  and  type(m3/2,integer)  and  type(m4/2,integer)
then
K12:=combinat[multinomial](4,4,0,0,0)*combinat[multinomial](n-
2,(m1-4)/2,m2/2,m3/2,m4/2); else K12:=0:fi:

>    if (m2-4)>=0 and type(m1/2,integer) and type((m2-4)/2,integer)
and      type(m3/2,integer)      and      type(m4/2,integer)      then
K22:=combinat[multinomial](4,0,4,0,0)*combinat[multinomial](n-
2,m1/2,(m2-4)/2,m3/2,m4/2); else K22:=0:fi:

>    if (m3-4)>=0 and type(m1/2,integer) and type(m2/2,integer) and
type((m3-4)/2,integer)         and         type(m4/2,integer)     then
K32:=combinat[multinomial](4,0,0,4,0)*combinat[multinomial](n-
2,m1/2,m2/2,(m3-4)/2,m4/2); else K32:=0:fi:

>    if (m4-4)>=0 and type(m1/2,integer) and type(m2/2,integer) and
```

```
type(m3/2,integer)          and          type((m4-4)/2,integer)          then
K32:=combinat[multinomial](4,0,0,0,4)*combinat[multinomial](n-
2,m1/2,m2/2,m3/2,(m4-4)/2); else K32:=0:fi:

>       if (m1-3)>=0 and (m2-1)>=0 and type((m1-3)/2,integer) and
type((m2-1)/2,integer)          and          type(m3/2,integer)          and
type(m4/2,integer)                                                        then
K42:=combinat[multinomial](4,3,1,0,0)*combinat[multinomial](n-
2,(m1-3)/2,(m2-1)/2,m3/2,m4/2); else K42:=0:fi:

>          if (m2-3)>=0 and (m3-1)>=0 and type(m1/2,integer) and
type((m2-3)/2,integer)          and          type((m3-1)/2,integer)          and
type(m4/2,integer)                                                        then
K52:=combinat[multinomial](4,0,3,1,0)*combinat[multinomial](n-
2,m1/2,(m2-3)/2,(m3-1)/2,m4/2); else K52:=0:fi:

>          if (m1-3)>=0 and (m3-1)>=0 and type((m1-3)/2,integer) and
type(m2/2,integer)          and          type((m3-1)/2,integer)          and
type(m4/2,integer)                                                        then
K62:=combinat[multinomial](4,3,0,1,0)*combinat[multinomial](n-
2,(m1-3)/2,m2/2,(m3-1)/2,m4/2); else K62:=0:fi:

>          if (m1-3)>=0 and (m4-1)>=0 and type((m1-3)/2,integer) and
type(m2/2,integer)          and          type((m4-1)/2,integer)          and
type(m3/2,integer)                                                        then
K72:=combinat[multinomial](4,3,0,0,1)*combinat[multinomial](n-
2,(m1-3)/2,m2/2,m3/2,(m4-1)/2); else K72:=0:fi:

>          if (m2-3)>=0 and (m4-1)>=0 and type(m1/2,integer) and
type((m2-3)/2,integer)          and          type((m4-1)/2,integer)          and
type(m3/2,integer)                                                        then
K822:=combinat[multinomial](4,0,3,0,1)*combinat[multinomial](n-
2,m1/2,(m2-3)/2,m3/2,(m4-1)/2); else K822:=0:fi:

> if (m3-3)>=0 and (m4-1)>=0 and type(m1/2,integer) and type((m3-
3)/2,integer)  and  type((m4-1)/2,integer)  and  type(m2/2,integer)
then
K82:=combinat[multinomial](4,0,0,3,1)*combinat[multinomial](n-
2,m1/2,m2/2,(m3-3)/2,(m4-1)/2); else K82:=0:fi:

> if    (m1-1)>=0   and   (m2-3)>=0   and   type((m1-1)/2,integer)   and
type((m2-3)/2,integer)          and          type(m3/2,integer)          and
type(m4/2,integer)                                                        then
K92:=combinat[multinomial](4,3,1,0,0)*combinat[multinomial](n-
2,(m1-1)/2,(m2-3)/2,m3/2,m4/2); else K92:=0:fi:

>          if (m2-1)>=0 and (m3-3)>=0 and type(m1/2,integer) and
type((m2-1)/2,integer)          and          type((m3-3)/2,integer)          and
type(m4/2,integer)                                                        then
K102:=combinat[multinomial](4,0,3,1,0)*combinat[multinomial](n-
2,m1/2,(m2-1)/2,(m3-3)/2,m4/2); else K102:=0:fi:

>          if (m1-1)>=0 and (m3-3)>=0 and type((m1-1)/2,integer) and
type(m2/2,integer)          and          type((m3-3)/2,integer)          and
type(m4/2,integer)                                                        then
```

```
K112:=combinat[multinomial](4,3,0,1,0)*combinat[multinomial](n-
2,(m1-1)/2,m2/2,(m3-3)/2,m4/2); else K112:=0:fi:

>       if (m1-1)>=0 and (m4-3)>=0 and type((m1-1)/2,integer) and
type(m2/2,integer)         and         type((m4-3)/2,integer)         and
type(m3/2,integer)                                                    then
K122:=combinat[multinomial](4,3,0,0,1)*combinat[multinomial](n-
2,(m1-1)/2,m2/2,m3/2,(m4-3)/2); else K122:=0:fi:

>         if  (m2-1)>=0  and   (m4-3)>=0  and  type(m1/2,integer)   and
type((m2-1)/2,integer)       and       type((m4-3)/2,integer)         and
type(m3/2,integer)                                                    then
K132:=combinat[multinomial](4,0,3,0,1)*combinat[multinomial](n-
2,m1/2,(m2-1)/2,m3/2,(m4-3)/2); else K132:=0:fi:

> if (m3-1)>=0 and (m4-3)>=0 and type(m1/2,integer) and type((m3-
1)/2,integer)  and  type((m4-3)/2,integer)  and  type(m2/2,integer)
then
K142:=combinat[multinomial](4,0,0,3,1)*combinat[multinomial](n-
2,m1/2,m2/2,(m3-1)/2,(m4-3)/2); else K142:=0:fi:

>       if  (m1-2)>=0  and   (m2-2)>=0  and  type((m1-2)/2,integer)  and
type((m2-2)/2,integer)        and        type(m3/2,integer)           and
type(m4/2,integer)                                                    then
K152:=combinat[multinomial](4,2,2,0,0)*combinat[multinomial](n-
2,(m1-2)/2,(m2-2)/2,m3/2,m4/2); else K152:=0:fi:

>         if  (m2-2)>=0  and   (m3-2)>=0  and  type(m1/2,integer)   and
type((m2-2)/2,integer)        and        type((m3-2)/2,integer)       and
type(m4/2,integer)                                                    then
K162:=combinat[multinomial](4,0,2,2,0)*combinat[multinomial](n-
2,m1/2,(m2-2)/2,(m3-2)/2,m4/2); else K162:=0:fi:

>       if  (m1-2)>=0  and   (m3-2)>=0  and  type((m1-2)/2,integer)  and
type(m2/2,integer)         and         type((m3-2)/2,integer)         and
type(m4/2,integer)                                                    then
K172:=combinat[multinomial](4,2,0,2,0)*combinat[multinomial](n-
2,(m1-2)/2,m2/2,(m3-2)/2,m4/2); else K172:=0:fi:

>         if  (m3-2)>=0  and   (m4-2)>=0  and  type(m1/2,integer)   and
type(m2/2,integer)      and     type((m3-2)/2,integer)     and    type((m4-
2)/2,integer)                                                         then
K182:=combinat[multinomial](4,0,0,2,2)*combinat[multinomial](n-
2,m1/2,m2/2,(m3-2)/2,(m4-2)/2); else K182:=0:fi:

>         if  (m1-2)>=0  and   (m4-2)>=0  and  type((m1-2)/2,integer) and
type(m2/2,integer)       and     type(m3/2,integer)      and     type((m4-
2)/2,integer)                                                         then
K192:=combinat[multinomial](4,2,0,0,2)*combinat[multinomial](n-
2,(m1-2)/2,m2/2,m3/2,(m4-2)/2); else K192:=0:fi:

>         if  (m2-2)>=0  and   (m4-2)>=0  and  type(m1/2,integer)   and
type((m2-2)/2,integer)        and        type(m3/2,integer)           and
type(m4/2,integer)                                                    then
K202:=combinat[multinomial](4,0,2,0,2)*combinat[multinomial](n-
```

```
2,m1/2,(m2-2)/2,m3/2,(m4-2)/2); else K202:=0:fi:

>        if  (m1-2)>=0  and  (m2-1)>=0  and  (m3-1)>=0  and  type((m1-
2)/2,integer)      and      type((m2-1)/2,integer)      and      type((m3-
1)/2,integer)      and      type(m4/2,      integer)      then
K212:=combinat[multinomial](4,2,1,1,0)*combinat[multinomial](n-
2,(m1-2)/2,(m2-1)/2,(m3-1)/2,m4/2); else K212:=0:fi:

> if  (m1-2)>=0  and  (m2-1)>=0  and  (m4-1)>=0  and  type((m1-
2)/2,integer)  and  type((m2-1)/2,integer)  and  type(m3/2,integer)
and             type((m4-1)/2,             integer)             then
K222:=combinat[multinomial](4,2,1,0,1)*combinat[multinomial](n-
2,(m1-2)/2,(m2-1)/2,m3/2,(m4-1)/2); else K222:=0:fi:

>        if  (m1-2)>=0  and  (m3-1)>=0  and  (m4-1)>=0  and  type((m1-
2)/2,integer)  and  type(m2/2,integer)  and  type((m3-1)/2,integer)
and             type((m4-1)/2,             integer)             then
K232:=combinat[multinomial](4,2,0,1,1)*combinat[multinomial](n-
2,(m1-2)/2,m2/2,(m3-1)/2,(m4-1)/2);  else  K232:=0:fi:if  (m2-2)>=0
and   (m3-1)>=0   and   (m4-1)>=0   and   type((m2-2)/2,integer)   and
type(m1/2,integer)  and  type((m3-1)/2,integer)  and  type((m4-1)/2,
integer)                                                         then
K2322:=combinat[multinomial](4,0,2,1,1)*combinat[multinomial](n-
2,(m2-2)/2,m1/2,(m3-1)/2,(m4-1)/2); else K2322:=0:fi:

> if  (m1-1)>=0  and  (m2-2)>=0  and  (m3-1)>=0  and  type((m1-
1)/2,integer)      and      type((m2-2)/2,integer)      and      type((m3-
1)/2,integer)      and      type(m4/2,      integer)      then
K242:=combinat[multinomial](4,1,2,1,0)*combinat[multinomial](n-
2,(m1-1)/2,(m2-2)/2,(m3-1)/2,m4/2); else K242:=0:fi:

> if  (m1-1)>=0  and  (m2-2)>=0  and  (m4-1)>=0  and  type((m1-
1)/2,integer)  and  type((m2-2)/2,integer)  and  type(m3/2,integer)
and             type((m4-1)/2,             integer)             then
K252:=combinat[multinomial](4,1,2,0,1)*combinat[multinomial](n-
2,(m1-1)/2,(m2-2)/2,m3/2,(m4-1)/2); else K252:=0:fi:

>        if  (m1-1)>=0  and  (m3-2)>=0  and  (m4-1)>=0  and  type((m1-
1)/2,integer)  and  type(m2/2,integer)  and  type((m3-2)/2,integer)
and             type((m4-1)/2,             integer)             then
K262:=combinat[multinomial](4,1,0,2,1)*combinat[multinomial](n-
2,(m1-1)/2,m2/2,(m3-2)/2,(m4-1)/2);  else  K262:=0:fi:if  (m2-1)>=0
and   (m3-2)>=0   and   (m4-1)>=0   and   type((m2-1)/2,integer)   and
type(m1/2,integer)  and  type((m3-2)/2,integer)  and  type((m4-1)/2,
integer)                                                         then
K2622:=combinat[multinomial](4,1,0,2,1)*combinat[multinomial](n-
2,(m2-1)/2,m1/2,(m3-2)/2,(m4-1)/2); else K2622:=0:fi:

> if  (m1-1)>=0  and  (m2-1)>=0  and  (m3-2)>=0  and  type((m1-
1)/2,integer)      and      type((m2-1)/2,integer)      and      type((m3-
2)/2,integer)      and      type(m4/2,      integer)      then
K272:=combinat[multinomial](4,1,2,1,0)*combinat[multinomial](n-
2,(m1-1)/2,(m2-1)/2,(m3-2)/2,m4/2); else K272:=0:fi:
```

83

```
> if    (m1-1)>=0   and    (m2-1)>=0   and   (m4-2)>=0   and   type((m1-
1)/2,integer)   and   type((m2-1)/2,integer)   and   type(m3/2,integer)
and            type((m4-2)/2,            integer)            then
K282:=combinat[multinomial](4,1,2,0,1)*combinat[multinomial](n-
2,(m1-1)/2,(m2-1)/2,m3/2,(m4-2)/2); else K282:=0:fi:

>        if   (m1-1)>=0   and   (m3-1)>=0   and   (m4-2)>=0   and type((m1-
1)/2,integer)   and   type(m2/2,integer)   and   type((m3-1)/2,integer)
and            type((m4-2)/2,            integer)            then
K292:=combinat[multinomial](4,1,0,2,1)*combinat[multinomial](n-
2,(m1-1)/2,m2/2,(m3-1)/2,(m4-2)/2);   else   K292:=0:fi:if   (m2-1)>=0
and    (m3-1)>=0   and    (m4-2)>=0   and   type((m2-1)/2,integer)   and
type(m1/2,integer)   and   type((m3-1)/2,integer)   and   type((m4-2)/2,
integer)                                                    then
K2922:=combinat[multinomial](4,0,1,2,1)*combinat[multinomial](n-
2,(m2-1)/2,m1/2,(m3-1)/2,(m4-2)/2); else K2922:=0:fi:

> if   (m1-1)>=0   and   (m2-1)>=0   and   (m3-1)>=0   and   (m4-1)>=0   and
type((m1-1)/2,integer)   and   type((m2-1)/2,integer)   and   type((m3-
1)/2,integer)          and          type((m4-1)/2,          integer)          then
K302:=combinat[multinomial](4,1,1,1,1)*combinat[multinomial](n-
2,(m1-1)/2,(m2-1)/2,(m3-1)/2,(m4-1)/2); else K302:=0:fi:

>
KT2:=K12+K22+K32+K42+K52+K62+K72+K82+K92+K102+K112+K122+K132+K142+
K152+K162+K172+K182+K192+K202+K212+K222+K232+K242+K252+K262+K272+K
282+K292+K302+K2322+K2622+K2922+K822:

>            if    type(m1/2,integer)    and    type(m2/2,integer)    and
type(m3/2,integer)          and          type(m4/2,integer)          then
P:=combinat[multinomial](n,m1/2,m2/2,m3/2,m4/2); else P:=0: fi:

> Ac(n,m1,m2,m3,m4):=(1/(4*n))*(S1+((n/2)-1)*P)-(1/8)*KT2:

> Aa(n,m1,m2,m3,m4):=(1/(2*n))*(S2+((n/2)+2)*P)+(1/4)*KT2:

> print (Ac(n,m1,m2,m3,m4),Aa(n,m1,m2,m3,m4));

> fi;

> end:

> end:
```

TO RUN THE PROGRAM: Type stereo(n,m1,m2,m3,m4); and press Enter to get the number of chiral and achiral skeletons of CnXm1Ym2Zm3Um4 respectively.

Exemple:

```
> stereo(6,9,1,1,1);
```
$$52, 6$$

```
> stereo(6,8,2,1,1);
```
$$240, 15$$

```
> stereo(6,7,3,1,1);
```
$$648, 24$$

```
> stereo(6,3,3,3,3);
```
$$15330, 144$$

```
> stereo(38,14,12,34,16);
```
$$4808971624142146035978501292019629440 0 \quad , 8398871645174010000$$

```
> stereo(12,9,3,6,6);
```
$$11452052430 , 96600$$

MAPLE PROGRAM FOR THE COMPUTATION OF THE NUMBERS OF CHIRAL AND ACHIRAL SKELETONS OF HETEROPOLYSUBSTITUTED MONOCYCLIC CYCLOALKANES CnXm1Ym2Zm3Um4Vm5

n = ring size, mi = number of non isomerizable substituents, i=1,2,3,4,5.

By Alphonse EMADAK

```
> restart:
```

Subprogram: To compute of the numbers of permutations induced by proper and improper symmetries of Dnh

```
> Ad1:=proc(x1)
> local y1;
> if type(x1,odd) or x1=2 then y1:=numtheory[phi](x1)
>            else   if    type(x1/2,odd)    and    x1<>2    then
y1:=2*numtheory[phi](x1)+numtheory[phi](x1/2);
>            else y1:=2*numtheory[phi](x1)
>       fi;
> fi;
> end:
```

Subprogram: To compute of the numbers of permutations induced by proper symmetries of Dnh

```
> Ad2:=proc(x2)
> local y2;
> if type(x2,odd) or type(x2,even) then y2:=numtheory[phi](x2);
> fi;
> end:
```

Subprogram: To compute the number Ac and Aa respectively of chiral and achiral graphs of heteropolysubstituted skeletons of cycloalkanes

```
>            stereo:=proc(n,m1,m2,m3,m4,m5)                    local
Dn,Dm1,Dm2,Dm3,Dm4,Dm5,Dc,S1,S2,P,KT1,KT2,K11,K21,K31,K41,K51,K61,
K71,K81,K91,K101,K111,K121,K131,K141,K151,K12,K22,K32,K42,K52,K62,
K72,K82,K92,K102,K112,K122,K132,K142,K152,K162,K172,K182,K192,K202
,K212,K222,K232,K242,K252,K262,K272,K282,K292,K302,K312,K322,K332,
K342,K352,K362,K372,K382,K392,K402,K412,K422,K432,K442,K452,K462,K
472,K482,K492,K502,K512,K522,K532,K542,K552,K562,K572,K582,K592,K6
02,K612,K622,K632,K642,K652,K662,K672,K682:
# CASE n ODD

>     if     type(n,odd)     then     Dn:=numtheory[divisors](2*n);
Dm1:=numtheory[divisors](m1):       Dm2:=numtheory[divisors](m2):
Dm3:=numtheory[divisors](m3):Dm4:=numtheory[divisors](m4):Dm5:=num
```

86

```
theory[divisors](m5):

> Dc:= convert((Dn intersect Dm1 intersect Dm2 intersect Dm3
intersect Dm4 intersect Dm5) minus {2}, list):

>                                        S1:=sum('(2*Ad2(Dc[i])-
Ad1(Dc[i]))*combinat[multinomial](2*n/Dc[i],m1/Dc[i],m2/Dc[i],m3/D
c[i],m4/Dc[i],m5/Dc[i])','i'=1..nops(Dc)):

>                                        S2:=sum('(Ad1(Dc[i])-
Ad2(Dc[i]))*combinat[multinomial](2*n/Dc[i],m1/Dc[i],m2/Dc[i],m3/D
c[i],m4/Dc[i],m5/Dc[i])','i'=1..nops(Dc)):

>    if (m1-2)>=0 and type((m1-2)/2,integer) and type(m2/2,integer)
and      type(m3/2,integer)      and      type(m4/2,integer)      and
type(m5/2,integer)                                              then
K11:=combinat[multinomial](2,2,0,0,0,0)*combinat[multinomial](n-
1,(m1-2)/2,m2/2,m3/2,m4/2,m5/2): else K11:=0:fi:

>    if (m2-2)>=0 and type(m1/2,integer) and type((m2-2)/2,integer)
and      type(m3/2,integer)      and      type(m4/2,integer)      then
K21:=combinat[multinomial](2,0,2,0,0,0)*combinat[multinomial](n-
1,m1/2,(m2-2)/2,m3/2,m4/2,m5/2): else K21:=0:fi:

>    if (m3-2)>=0 and     type(m1/2,integer) and type(m2/2,integer)
and      type((m3-2)/2,integer)      and      type(m4/2,integer)      and
type(m5/2,integer)                                              then
K31:=combinat[multinomial](2,0,0,2,0,0)*combinat[multinomial](n-
1,m1/2,m2/2,(m3-2)/2, m4/2,m5/2): else K31:=0:fi:

>    if (m4-2)>=0 and     type(m1/2,integer) and type(m2/2,integer)
and      type(m3/2,integer)      and      type((m4-2)/2,integer)      and
type(m5/2,integer)                                              then
K41:=combinat[multinomial](2,0,0,0,2,0)*combinat[multinomial](n-
1,m1/2,m2/2,(m3-2)/2, m4/2,m5/2): else K41:=0:fi:

>    if (m5-2)>=0 and     type(m1/2,integer) and type(m2/2,integer)
and      type(m3/2,integer)      and      type((m5-2)/2,integer)      and
type(m4/2,integer)                                              then
K51:=combinat[multinomial](2,0,0,0,0,2)*combinat[multinomial](n-
1,m1/2,m2/2,(m5-2)/2, m4/2,m3/2): else K51:=0:fi:

>        if (m1-1)>=0 and (m2-1)>=0 and type((m1-1)/2,integer) and
type((m2-1)/2,integer)      and      type(m3/2,integer)      and
type(m4/2,integer)      and      type(m5/2,integer)      then
K61:=combinat[multinomial](2,1,1,0,0,0)*combinat[multinomial](n-
1,(m1-1)/2,(m2-1)/2,m3/2,m4/2,m5/2): else K61:=0:fi:

>        if (m1-1)>=0 and (m3-1)>=0 and type((m1-1)/2,integer) and
type(m2/2,integer)      and      type((m3-1)/2,integer)      and
type(m4/2,integer)      and      type(m5/2,integer)      then
K71:=combinat[multinomial](2,1,0,1,0,0)*combinat[multinomial](n-
1,(m1-1)/2,m2/2,(m3-1)/2,m4/2,m5/2): else K71:=0:fi:

>        if (m2-1)>=0 and (m3-1)>=0 and type(m1/2,integer) and
type((m2-1)/2,integer)      and      type((m3-1)/2,integer)      and
```

```
type(m4/2,integer)            and            type(m5/2,integer)            then
K81:=combinat[multinomial](2,0,1,1,0,0)*combinat[multinomial](n-
1,m1/2,(m2-1)/2,(m3-1)/2,m4/2,m5/2): else K81:=0:fi:

>        if  (m1-1)>=0  and  (m4-1)>=0  and  type((m1-1)/2,integer)  and
type(m2/2,integer)            and            type((m4-1)/2,integer)            and
type(m3/2,integer)            and            type(m5/2,integer)            then
K91:=combinat[multinomial](2,1,0,0,1,0)*combinat[multinomial](n-
1,(m1-1)/2,m2/2,m3/2,(m4-1)/2,m5/2): else K91:=0:fi:

>          if  (m2-1)>=0  and  (m4-1)>=0  and  type(m1/2,integer)  and
type((m2-1)/2,integer)        and          type((m4-1)/2,integer)         and
type(m3/2,integer)            and            type(m5/2,integer)            then
K101:=combinat[multinomial](2,0,1,0,1,0)*combinat[multinomial](n-
1,m1/2,(m2-1)/2,m3/2,(m4-1)/2,m5/2): else K101:=0:fi:

>          if  (m3-1)>=0  and  (m4-1)>=0  and  type(m1/2,integer)  and
type((m3-1)/2,integer)        and          type((m4-1)/2,integer)         and
type(m2/2,integer)            and            type(m5/2,integer)            then
K111:=combinat[multinomial](2,0,0,1,1,0)*combinat[multinomial](n-
1,m1/2,(m3-1)/2,m2/2,(m4-1)/2,m5/2): else K111:=0:fi:

>       if  (m1-1)>=0  and  (m5-1)>=0  and  type((m1-1)/2,integer)  and
type(m3/2,integer)  and  type(m4/2,integer)  and  type(m2/2,integer)
and                     type((m5-1)/2,integer)                     then
K121:=combinat[multinomial](2,1,0,0,0,1)*combinat[multinomial](n-
1,(m1-1)/2,m2/2,m3/2,m4/2,(m5-1)/2): else K121:=0:fi:

>  if    (m2-1)>=0   and    (m5-1)>=0   and    type(m1/2,integer)   and
type(m3/2,integer)         and        type(m4/2,integer)        and       type((m2-
1)/2,integer)            and            type((m5-1)/2,integer)            then
K131:=combinat[multinomial](2,0,1,0,0,1)*combinat[multinomial](n-
1,m1/2,(m2-1)/2,m3/2,m4/2,(m5-1)/2): else K131:=0:fi:

> if  (m3-1)>=0  and  (m5-1)>=0  and  type(m1/2,integer)  and  type((m3-
1)/2,integer)  and  type(m4/2,integer)  and  type(m2/2,integer)  and
type((m5-1)/2,integer)                                               then
K141:=combinat[multinomial](2,0,0,1,0,1)*combinat[multinomial](n-
1,m1/2,m2/2,(m3-1)/2,m4/2,(m5-1)/2): else K141:=0:fi:

>  if    (m4-1)>=0   and    (m5-1)>=0   and    type((m4-1)/2,integer)   and
type(m3/2,integer)  and  type(m1/2,integer)  and  type(m2/2,integer)
and                     type((m5-1)/2,integer)                     then
K151:=combinat[multinomial](2,0,0,0,1,1)*combinat[multinomial](n-
1,(m4-1)/2,m2/2,m3/2,m1/2,(m5-1)/2): else K151:=0:fi:

>
KT1:=K11+K21+K31+K41+K51+K61+K71+K81+K91+K101+K111+K121+K131+K141+
K151:

>          if   type(m1/2,integer)   and   type(m2/2,integer)   and
type(m3/2,integer)  and  type(m4/2,integer)  and  type(m5/2,integer)
then     P:=combinat[multinomial](n,m1/2,m2/2,m3/2,m4/2,m5/2):     else
P:=0: fi:
```

```
> Ac(n,m1,m2,m3,m4,m5):=(1/(4*n))*(S1+(n-1)*P)-(1/4)*KT1:

> Aa(n,m1,m2,m3,m4,m5):=(1/(2*n))*(S2+P)+(1/2)*KT1:

> print (Ac(n,m1,m2,m3,m4,m5),Aa(n,m1,m2,m3,m4,m5));

>                                                           else
# CASE n EVEN

>     if     type(n,even)     then     Dn:=numtheory[divisors](n);
Dm1:=numtheory[divisors](m1):         Dm2:=numtheory[divisors](m2):
Dm3:=numtheory[divisors](m3):Dm4:=numtheory[divisors](m4):Dm5:=num
theory[divisors](m5):

> Dc:= convert((Dn intersect Dm1 intersect Dm2 intersect Dm3
intersect Dm4 intersect Dm5) minus {2}, list):

>                                             S1:=sum('(2*Ad2(Dc[i])-
Ad1(Dc[i]))*combinat[multinomial](2*n/Dc[i],m1/Dc[i],m2/Dc[i],m3/D
c[i],m4/Dc[i],m5/Dc[i])','i'=1..nops(Dc)):

>                                             S2:=sum('(Ad1(Dc[i])-
Ad2(Dc[i]))*combinat[multinomial](2*n/Dc[i],m1/Dc[i],m2/Dc[i],m3/D
c[i],m4/Dc[i],m5/Dc[i])','i'=1..nops(Dc)):

>     if (m1-4)>=0 and type((m1-4)/2,integer) and type(m2/2,integer)
and        type(m3/2,integer)        and        type(m4/2,integer)        and
type(m5/2,integer)                                                  then
K12:=combinat[multinomial](4,4,0,0,0,0)*combinat[multinomial](n-
2,(m1-4)/2,m2/2,m3/2,m4/2,m5/2); else K12:=0:fi:

>     if (m2-4)>=0 and type(m1/2,integer) and type((m2-4)/2,integer)
and        type(m3/2,integer)        and        type(m4/2,integer)        and
type(m5/2,integer)                                                  then
K22:=combinat[multinomial](4,0,4,0,0,0)*combinat[multinomial](n-
2,m1/2,(m2-4)/2,m3/2,m4/2,m5/2); else K22:=0:fi:

>     if (m3-4)>=0 and type(m1/2,integer) and type(m2/2,integer) and
type((m3-4)/2,integer)        and        type(m4/2,integer)        and
type(m5/2,integer)                                                  then
K32:=combinat[multinomial](4,0,0,4,0,0)*combinat[multinomial](n-
2,m1/2,m2/2,(m3-4)/2,m4/2,m5/2); else K32:=0:fi:

>     if (m4-4)>=0 and type(m1/2,integer) and type(m2/2,integer) and
type(m3/2,integer)        and        type((m4-4)/2,integer)        and
type(m5/2,integer)                                                  then
K32:=combinat[multinomial](4,0,0,0,4,0)*combinat[multinomial](n-
2,m1/2,m2/2,m3/2,(m4-4)/2,m5/2); else K32:=0:fi:

>     if (m5-4)>=0 and type(m1/2,integer) and type(m2/2,integer) and
type(m3/2,integer)     and     type(m4/2,integer)     and     type((m5-
1)/2,integer)                                                       then
K42:=combinat[multinomial](4,0,0,0,0,4)*combinat[multinomial](n-
2,m1/2,m2/2,m3/2,(m5-4)/2,m4/2); else K42:=0:fi:

> if   (m1-3)>=0   and   (m2-1)>=0   and   type((m1-3)/2,integer)   and
type((m2-1)/2,integer)          and          type(m3/2,integer)          and
```

89

```
type(m4/2,integer)          and          type(m5/2,integer)          then
K52:=combinat[multinomial](4,3,1,0,0,0)*combinat[multinomial](n-
2,(m1-3)/2,(m2-1)/2,m3/2,m4/2,m5/2); else K52:=0:fi:

>        if  (m2-3)>=0  and  (m3-1)>=0  and  type(m1/2,integer)  and
type((m2-3)/2,integer)          and          type((m3-1)/2,integer)          and
type(m4/2,integer)          and          type(m5/2,integer)          then
K62:=combinat[multinomial](4,0,3,1,0,0)*combinat[multinomial](n-
2,m1/2,(m2-3)/2,(m3-1)/2,m4/2,m5/2); else K62:=0:fi:

>        if  (m1-3)>=0  and  (m3-1)>=0  and  type((m1-3)/2,integer)  and
type(m2/2,integer)          and          type((m3-1)/2,integer)          and
type(m4/2,integer)          and          type(m5/2,integer)          then
K72:=combinat[multinomial](4,3,0,1,0,0)*combinat[multinomial](n-
2,(m1-3)/2,m2/2,(m3-1)/2,m4/2,m5/2); else K72:=0:fi:

>        if  (m1-3)>=0  and  (m4-1)>=0  and  type((m1-3)/2,integer)  and
type(m2/2,integer)          and          type((m4-1)/2,integer)          and
type(m3/2,integer)          and          type(m5/2,integer)          then
K82:=combinat[multinomial](4,3,0,0,1,0)*combinat[multinomial](n-
2,(m1-3)/2,m2/2,m3/2,(m4-1)/2,m5/2); else K82:=0:fi:

>        if  (m2-3)>=0  and  (m4-1)>=0  and  type(m1/2,integer)  and
type((m2-3)/2,integer)          and          type((m4-1)/2,integer)          and
type(m3/2,integer)          and          type(m5/2,integer)          then
K92:=combinat[multinomial](4,0,3,0,1,0)*combinat[multinomial](n-
2,m1/2,(m2-3)/2,m3/2,(m4-1)/2,m5/2); else K92:=0:fi:

>        if  (m3-3)>=0  and  (m4-1)>=0  and  type(m1/2,integer)  and
type((m3-3)/2,integer)          and          type((m4-1)/2,integer)          and
type(m2/2,integer)          and          type(m5/2,integer)          then
K102:=combinat[multinomial](4,0,0,3,1,0)*combinat[multinomial](n-
2,m1/2,m2/2,(m3-3)/2,(m4-1)/2,m5/2); else K102:=0:fi:

>        if (m1-3)>=0 and (m5-1)>=0 and type((m1-3)/2,integer) and
type(m2/2,integer)  and  type(m3/2,integer)  and  type(m4/2,integer)
and                    type((m5-1)/2,integer)                    then
K112:=combinat[multinomial](4,3,0,1,0,0)*combinat[multinomial](n-
2,(m1-3)/2,m2/2,m3/2,m4/2,(m5-1)/2); else K112:=0:fi:

>        if (m2-3)>=0 and (m5-1)>=0 and type((m2-3)/2,integer) and
type(m1/2,integer)  and  type(m3/2,integer)  and  type(m4/2,integer)
and                    type((m5-1)/2,integer)                    then
K122:=combinat[multinomial](4,0,3,0,0,1)*combinat[multinomial](n-
2,(m2-3)/2,m1/2,m3/2,m4/2,(m5-1)/2); else K122:=0:fi:

>        if (m3-3)>=0 and (m5-1)>=0 and type((m3-3)/2,integer) and
type(m2/2,integer)  and  type(m1/2,integer)  and  type(m4/2,integer)
and                    type((m5-1)/2,integer)                    then
K132:=combinat[multinomial](4,0,0,3,0,1)*combinat[multinomial](n-
2,(m3-3)/2,m2/2,m1/2,m4/2,(m5-1)/2); else K132:=0:fi:

> if  (m4-3)>=0  and  (m5-1)>=0  and  type((m4-3)/2,integer)  and
type(m2/2,integer)  and  type(m3/2,integer)  and  type(m1/2,integer)
and                    type((m5-1)/2,integer)                    then
```

```
K142:=combinat[multinomial](4,0,0,0,3,1)*combinat[multinomial](n-
2,(m4-3)/2,m2/2,m3/2,m1/2,(m5-1)/2); else K142:=0:fi:

> if (m1-1)>=0 and (m2-3)>=0 and type((m1-1)/2,integer) and
type((m2-3)/2,integer)        and        type(m3/2,integer)        and
type(m4/2,integer)        and        type(m5/2,integer)        then
K152:=combinat[multinomial](4,1,3,0,0,0)*combinat[multinomial](n-
2,(m1-1)/2,(m2-3)/2,m3/2,m4/2,m5/2); else K152:=0:fi:

>        if (m2-1)>=0 and (m3-3)>=0 and type(m1/2,integer) and
type((m2-1)/2,integer)        and        type((m3-3)/2,integer)        and
type(m4/2,integer)        and        type(m5/2,integer)        then
K162:=combinat[multinomial](4,0,1,3,0,0)*combinat[multinomial](n-
2,m1/2,(m2-1)/2,(m3-3)/2,m4/2,m5/2); else K162:=0:fi:

>        if (m1-1)>=0 and (m3-3)>=0 and type((m1-1)/2,integer) and
type(m2/2,integer)        and        type((m3-3)/2,integer)        and
type(m4/2,integer)        and        type(m5/2,integer)        then
K172:=combinat[multinomial](4,1,0,3,0,0)*combinat[multinomial](n-
2,(m1-1)/2,m2/2,(m3-3)/2,m4/2,m5/2); else K172:=0:fi:

>        if (m1-1)>=0 and (m4-3)>=0 and type((m1-1)/2,integer) and
type(m2/2,integer)        and        type((m4-3)/2,integer)        and
type(m3/2,integer)        and        type(m5/2,integer)        then
K182:=combinat[multinomial](4,1,0,0,3,0)*combinat[multinomial](n-
2,(m1-1)/2,m2/2,m3/2,(m4-3)/2,m5/2); else K182:=0:fi:

>        if (m2-1)>=0 and (m4-3)>=0 and type(m1/2,integer) and
type((m2-1)/2,integer)        and        type((m4-3)/2,integer)        and
type(m3/2,integer)        and        type(m5/2,integer)        then
K192:=combinat[multinomial](4,0,1,0,3,0)*combinat[multinomial](n-
2,m1/2,(m2-1)/2,m3/2,(m4-3)/2,m5/2); else K192:=0:fi:

>        if (m3-1)>=0 and (m4-3)>=0 and type(m1/2,integer) and
type((m3-1)/2,integer)        and        type((m4-3)/2,integer)        and
type(m2/2,integer)        and        type(m5/2,integer)        then
K202:=combinat[multinomial](4,0,0,1,3,0)*combinat[multinomial](n-
2,m1/2,m2/2,(m3-1)/2,(m4-3)/2,m5/2); else K202:=0:fi:

>        if (m1-1)>=0 and (m5-3)>=0 and type((m1-1)/2,integer) and
type(m2/2,integer)  and  type(m3/2,integer)  and  type(m4/2,integer)
and                type((m5-3)/2,integer)                then
K212:=combinat[multinomial](4,1,0,0,3,0)*combinat[multinomial](n-
2,(m1-1)/2,m2/2,m3/2,m4/2,(m5-3)/2); else K212:=0:fi:

>        if (m2-1)>=0 and (m5-3)>=0 and type((m2-1)/2,integer) and
type(m1/2,integer)  and  type(m3/2,integer)  and  type(m4/2,integer)
and                type((m5-3)/2,integer)                then
K222:=combinat[multinomial](4,0,1,0,0,3)*combinat[multinomial](n-
2,(m2-1)/2,m1/2,m3/2,m4/2,(m5-3)/2); else K222:=0:fi:

>        if (m3-1)>=0 and (m5-3)>=0 and type((m3-1)/2,integer) and
type(m2/2,integer)  and  type(m1/2,integer)  and  type(m4/2,integer)
and                type((m5-3)/2,integer)                then
K232:=combinat[multinomial](4,0,0,1,0,3)*combinat[multinomial](n-
```

```
2,(m3-1)/2,m2/2,m1/2,m4/2,(m5-3)/2); else K232:=0:fi:

>        if (m4-1)>=0 and (m5-3)>=0 and type((m4-1)/2,integer) and
type(m2/2,integer) and type(m3/2,integer) and type(m1/2,integer)
and                    type((m5-3)/2,integer)                    then
K242:=combinat[multinomial](4,0,0,0,1,3)*combinat[multinomial](n-
2,(m4-1)/2,m2/2,m3/2,m1/2,(m5-3)/2); else K242:=0:fi:

> if  (m1-2)>=0 and  (m2-2)>=0 and  type((m1-2)/2,integer)  and
type((m2-2)/2,integer)      and      type(m3/2,integer)       and
type(m4/2,integer)        and        type(m5/2,integer)       then
K252:=combinat[multinomial](4,2,2,0,0,0)*combinat[multinomial](n-
2,(m1-2)/2,(m2-2)/2,m3/2,m4/2,m5/2); else K252:=0:fi:

>        if (m2-2)>=0 and  (m3-2)>=0 and  type(m1/2,integer)  and
type((m2-2)/2,integer)      and      type((m3-2)/2,integer)    and
type(m4/2,integer)        and        type(m5/2,integer)        then
K262:=combinat[multinomial](4,0,2,2,0,0)*combinat[multinomial](n-
2,m1/2,(m2-2)/2,(m3-2)/2,m4/2,m5/2); else K262:=0:fi:

>        if (m1-2)>=0 and  (m3-2)>=0 and  type((m1-2)/2,integer) and
type(m2/2,integer)          and          type((m3-2)/2,integer)  and
type(m4/2,integer)                                              then
K272:=combinat[multinomial](4,2,0,2,0)*combinat[multinomial](n-
2,(m1-2)/2,m2/2,(m3-2)/2,m4/2,m5/2); else K272:=0:fi:

>        if (m3-2)>=0 and  (m4-2)>=0 and  type(m1/2,integer)  and
type(m2/2,integer) and type((m3-2)/2,integer) and type((m4-
2)/2,integer)          and          type(m5/2,integer)         then
K282:=combinat[multinomial](4,0,0,2,2,0)*combinat[multinomial](n-
2,m1/2,m2/2,(m3-2)/2,(m4-2)/2,m5/2); else K282:=0:fi:

>        if (m1-2)>=0 and  (m4-2)>=0 and  type((m1-2)/2,integer) and
type(m2/2,integer) and type(m3/2,integer) and type((m4-
2)/2,integer)          and          type(m5/2,integer)         then
K292:=combinat[multinomial](4,2,0,0,2,0)*combinat[multinomial](n-
2,(m1-2)/2,m2/2,m3/2,(m4-2)/2,m5/2); else K292:=0:fi:

>        if (m2-2)>=0 and  (m4-2)>=0 and  type(m1/2,integer)  and
type((m2-2)/2,integer)      and      type(m3/2,integer)        and
type(m4/2,integer)        and        type(m5/2,integer)        then
K302:=combinat[multinomial](4,0,2,0,2,0)*combinat[multinomial](n-
2,m1/2,(m2-2)/2,m3/2,(m4-2)/2,m5/2); else K302:=0:fi:

>        if (m1-2)>=0 and  (m5-2)>=0 and  type((m1-2)/2,integer) and
type(m2/2,integer) and type(m3/2,integer) and type((m5-
2)/2,integer)          and          type(m4/2,integer)         then
K312:=combinat[multinomial](4,2,0,0,0,2)*combinat[multinomial](n-
2,(m1-2)/2,m2/2,m3/2,m4/2,(m5-2)/2); else K312:=0:fi:

>        if (m2-2)>=0 and  (m5-2)>=0 and  type((m2-2)/2,integer) and
type(m1/2,integer) and type(m3/2,integer) and type((m5-
2)/2,integer)          and          type(m4/2,integer)         then
K312:=combinat[multinomial](4,0,2,0,0,2)*combinat[multinomial](n-
2,(m2-2)/2,m1/2,m3/2,m4/2,(m5-2)/2); else K312:=0:fi:
```

92

```
>        if  (m3-2)>=0  and  (m5-2)>=0  and  type((m3-2)/2,integer)  and
type(m1/2,integer)      and      type(m2/2,integer)      and      type((m5-
2)/2,integer)            and            type(m4/2,integer)            then
K322:=combinat[multinomial](4,0,0,2,0,2)*combinat[multinomial](n-
2,(m3-2)/2,m1/2,m2/2,m4/2,(m5-2)/2); else K322:=0:fi:

> if  (m4-2)>=0  and  (m5-2)>=0  and  type((m4-2)/2,integer)  and
type(m1/2,integer)      and      type(m3/2,integer)      and      type((m5-
2)/2,integer)            and            type(m2/2,integer)            then
K332:=combinat[multinomial](4,0,0,0,2,2)*combinat[multinomial](n-
2,(m4-2)/2,m1/2,m3/2,m2/2,(m5-2)/2); else K332:=0:fi:

>        if  (m1-2)>=0  and  (m2-1)>=0  and  (m3-1)>=0  and  type((m1-
2)/2,integer)      and      type((m2-1)/2,integer)      and      type((m3-
1)/2,integer) and type(m4/2,integer) and type(m5/2,integer) then
K342:=combinat[multinomial](4,2,1,1,0,0)*combinat[multinomial](n-
2,(m1-2)/2,(m2-1)/2,(m3-1)/2,m4/2,m5/2); else K342:=0:fi:

>        if  (m1-2)>=0  and  (m2-1)>=0  and  (m4-1)>=0  and  type((m1-
2)/2,integer)  and  type((m2-1)/2,integer)  and  type(m3/2,integer)
and    type((m4-1)/2,    integer)    and    type(m5/2,integer)    then
K352:=combinat[multinomial](4,2,1,0,1,0)*combinat[multinomial](n-
2,(m1-2)/2,(m2-1)/2,m3/2,(m4-1)/2,m5/2); else K352:=0:fi:

>        if  (m1-2)>=0  and  (m3-1)>=0  and  (m4-1)>=0  and  type((m1-
2)/2,integer)  and  type(m2/2,integer)  and  type((m3-1)/2,integer)
and    type((m4-1)/2,    integer)    and    type(m5/2,integer)    then
K362:=combinat[multinomial](4,2,0,1,1,0)*combinat[multinomial](n-
2,(m1-2)/2,m2/2,(m3-1)/2,(m4-1)/2,m5/2); else K362:=0:fi:

>        if  (m1-2)>=0  and  (m2-1)>=0  and  (m5-1)>=0  and  type((m1-
2)/2,integer)      and      type((m2-1)/2,integer)      and      type((m5-
1)/2,integer) and type(m4/2, integer) and type(m3/2,integer) then
K372:=combinat[multinomial](4,2,1,0,0,1)*combinat[multinomial](n-
2,(m1-2)/2,(m2-1)/2,(m5-1)/2,m4/2,m3/2); else K372:=0:fi:

>        if  (m1-2)>=0  and  (m3-1)>=0  and  (m5-1)>=0  and  type((m1-
2)/2,integer)  and  type(m2/2,integer)  and  type((m3-1)/2,integer)
and    type((m5-1)/2,    integer)    and    type(m4/2,integer)    then
K382:=combinat[multinomial](4,2,0,1,0,1)*combinat[multinomial](n-
2,(m1-2)/2,m2/2,(m3-1)/2,(m5-1)/2,m4/2); else K382:=0:fi:

>        if  (m1-2)>=0  and  (m5-1)>=0  and  (m4-1)>=0  and  type((m1-
2)/2,integer)  and  type(m2/2,integer)  and  type((m5-1)/2,integer)
and    type((m4-1)/2,    integer)    and    type(m3/2,integer)    then
K392:=combinat[multinomial](4,2,0,1,1,0)*combinat[multinomial](n-
2,(m1-2)/2,m2/2,(m5-1)/2,(m4-1)/2,m3/2); else K392:=0:fi:

> if  (m2-2)>=0  and  (m3-1)>=0  and  (m4-1)>=0  and  type((m2-
2)/2,integer)  and  type(m1/2,integer)  and  type((m3-1)/2,integer)
and    type((m4-1)/2,    integer)    and    type(m5/2,integer)    then
K402:=combinat[multinomial](4,2,0,1,1,0)*combinat[multinomial](n-
2,(m2-2)/2,m1/2,(m3-1)/2,(m4-1)/2,m5/2); else K402:=0:fi:

>        if  (m2-2)>=0  and  (m3-1)>=0  and  (m5-1)>=0  and  type((m2-
```

```
2)/2,integer)      and      type((m3-1)/2,integer)      and      type((m5-
1)/2,integer) and type(m1/2, integer) and type(m4/2,integer) then
K412:=combinat[multinomial](4,2,1,0,0,1)*combinat[multinomial](n-
2,(m2-2)/2,(m3-1)/2,(m5-1)/2,m4/2,m1/2); else K412:=0:fi:

>        if  (m2-2)>=0  and  (m4-1)>=0  and  (m5-1)>=0  and  type((m2-
2)/2,integer)  and  type(m1/2,integer)  and  type((m4-1)/2,integer)
and    type((m5-1)/2,    integer)    and    type(m3/2,integer)    then
K422:=combinat[multinomial](4,2,0,1,1,0)*combinat[multinomial](n-
2,(m2-2)/2,m1/2,(m4-1)/2,(m5-1)/2,m3/2); else K422:=0:fi:

>        if  (m3-2)>=0  and  (m5-1)>=0  and  (m4-1)>=0  and  type((m3-
2)/2,integer)  and  type(m2/2,integer)  and  type((m5-1)/2,integer)
and    type((m4-1)/2,    integer)    and    type(m1/2,integer)    then
K432:=combinat[multinomial](4,2,0,1,1,0)*combinat[multinomial](n-
2,(m3-2)/2,m2/2,(m5-1)/2,(m4-1)/2,m1/2); else K432:=0:fi:

> if   (m1-1)>=0   and   (m2-2)>=0   and   (m3-1)>=0   and   type((m1-
1)/2,integer)      and      type((m2-2)/2,integer)      and      type((m3-
1)/2,integer) and type(m4/2, integer) and type(m5/2,integer) then
K442:=combinat[multinomial](4,2,1,1,0,0)*combinat[multinomial](n-
2,(m1-1)/2,(m2-2)/2,(m3-1)/2,m4/2,m5/2); else K442:=0:fi:

>        if  (m1-1)>=0  and  (m2-2)>=0  and  (m4-1)>=0  and  type((m1-
1)/2,integer)  and  type((m2-2)/2,integer)  and  type(m3/2,integer)
and    type((m4-1)/2,    integer)    and    type(m5/2,integer)    then
K452:=combinat[multinomial](4,2,1,0,1,0)*combinat[multinomial](n-
2,(m1-1)/2,(m2-2)/2,m3/2,(m4-1)/2,m5/2); else K452:=0:fi:

>        if  (m1-1)>=0  and  (m3-2)>=0  and  (m4-1)>=0  and  type((m1-
1)/2,integer)  and  type(m2/2,integer)  and  type((m3-2)/2,integer)
and    type((m4-1)/2,    integer)    and    type(m5/2,integer)    then
K462:=combinat[multinomial](4,2,0,1,1,0)*combinat[multinomial](n-
2,(m1-1)/2,m2/2,(m3-2)/2,(m4-1)/2,m5/2); else K462:=0:fi:

>        if  (m1-1)>=0  and  (m2-2)>=0  and  (m5-1)>=0  and  type((m1-
1)/2,integer)      and      type((m2-2)/2,integer)      and      type((m5-
1)/2,integer) and type(m4/2, integer) and type(m3/2,integer) then
K472:=combinat[multinomial](4,2,1,0,0,1)*combinat[multinomial](n-
2,(m1-1)/2,(m2-2)/2,(m5-1)/2,m4/2,m3/2); else K472:=0:fi:

>        if  (m1-1)>=0  and  (m3-2)>=0  and  (m5-1)>=0  and  type((m1-
1)/2,integer)  and  type(m2/2,integer)  and  type((m3-2)/2,integer)
and    type((m5-1)/2,    integer)    and    type(m4/2,integer)    then
K482:=combinat[multinomial](4,2,0,1,1,0)*combinat[multinomial](n-
2,(m1-1)/2,m2/2,(m3-2)/2,(m5-1)/2,m4/2); else K482:=0:fi:

>        if  (m1-1)>=0  and  (m5-1)>=0  and  (m4-2)>=0  and  type((m1-
1)/2,integer)  and  type(m2/2,integer)  and  type((m5-1)/2,integer)
and    type((m4-2)/2,    integer)    and    type(m3/2,integer)    then
K492:=combinat[multinomial](4,2,0,1,1,0)*combinat[multinomial](n-
2,(m1-1)/2,m2/2,(m5-1)/2,(m4-2)/2,m3/2); else K492:=0:fi:

> if   (m2-1)>=0   and   (m3-2)>=0   and   (m4-1)>=0   and   type((m2-
1)/2,integer)  and  type(m1/2,integer)  and  type((m3-2)/2,integer)
```

94

```
and    type((m4-1)/2,    integer)    and    type(m5/2,integer)    then
K502:=combinat[multinomial](4,2,0,1,1,0)*combinat[multinomial](n-
2,(m2-2)/2,m1/2,(m3-1)/2,(m4-1)/2,m5/2); else K502:=0:fi:

>        if  (m2-1)>=0  and  (m3-2)>=0  and  (m5-1)>=0  and  type((m2-
1)/2,integer)      and      type((m3-2)/2,integer)      and      type((m5-
1)/2,integer) and type(m1/2, integer) and type(m4/2,integer) then
K512:=combinat[multinomial](4,2,1,0,0,1)*combinat[multinomial](n-
2,(m2-1)/2,(m3-2)/2,(m5-1)/2,m4/2,m1/2); else K512:=0:fi:

>        if  (m2-1)>=0  and  (m4-2)>=0  and  (m5-1)>=0  and  type((m2-
1)/2,integer)  and  type(m1/2,integer)  and  type((m4-2)/2,integer)
and    type((m5-1)/2,    integer)    and    type(m3/2,integer)    then
K522:=combinat[multinomial](4,2,0,1,1,0)*combinat[multinomial](n-
2,(m2-1)/2,m1/2,(m4-2)/2,(m5-1)/2,m3/2); else K522:=0:fi:

>        if  (m3-1)>=0  and  (m5-1)>=0  and  (m4-2)>=0  and  type((m3-
1)/2,integer)  and  type(m2/2,integer)  and  type((m5-1)/2,integer)
and    type((m4-2)/2,    integer)    and    type(m1/2,integer)    then
K532:=combinat[multinomial](4,2,0,1,1,0)*combinat[multinomial](n-
2,(m3-1)/2,m2/2,(m5-1)/2,(m4-2)/2,m1/2); else K532:=0:fi:

> if  (m1-1)>=0  and  (m2-1)>=0  and  (m3-2)>=0  and  type((m1-
1)/2,integer)      and      type((m2-1)/2,integer)      and      type((m3-
2)/2,integer) and type(m4/2, integer) and type(m5/2,integer) then
K542:=combinat[multinomial](4,2,1,1,0,0)*combinat[multinomial](n-
2,(m1-1)/2,(m2-1)/2,(m3-2)/2,m4/2,m5/2); else K542:=0:fi:

>        if  (m1-1)>=0  and  (m2-1)>=0  and  (m4-2)>=0  and  type((m1-
1)/2,integer)  and  type((m2-1)/2,integer)  and  type(m3/2,integer)
and    type((m4-2)/2,    integer)    and    type(m5/2,integer)    then
K552:=combinat[multinomial](4,2,1,0,1,0)*combinat[multinomial](n-
2,(m1-1)/2,(m2-1)/2,m3/2,(m4-2)/2,m5/2); else K552:=0:fi:

>        if  (m1-1)>=0  and  (m3-1)>=0  and  (m4-2)>=0  and  type((m1-
1)/2,integer)  and  type(m2/2,integer)  and  type((m3-1)/2,integer)
and    type((m4-2)/2,    integer)    and    type(m5/2,integer)    then
K562:=combinat[multinomial](4,2,0,1,1,0)*combinat[multinomial](n-
2,(m1-1)/2,m2/2,(m3-1)/2,(m4-2)/2,m5/2); else K562:=0:fi:

>        if  (m1-1)>=0  and  (m2-1)>=0  and  (m5-2)>=0  and  type((m1-
1)/2,integer)      and      type((m2-1)/2,integer)      and      type((m5-
2)/2,integer) and type(m4/2, integer) and type(m3/2,integer) then
K572:=combinat[multinomial](4,2,1,0,0,1)*combinat[multinomial](n-
2,(m1-1)/2,(m2-1)/2,(m5-2)/2,m4/2,m3/2); else K572:=0:fi:

>        if  (m1-1)>=0  and  (m3-1)>=0  and  (m5-2)>=0  and  type((m1-
1)/2,integer)  and  type(m2/2,integer)  and  type((m3-1)/2,integer)
and    type((m5-2)/2,    integer)    and    type(m4/2,integer)    then
K582:=combinat[multinomial](4,2,0,1,1,0)*combinat[multinomial](n-
2,(m1-1)/2,m2/2,(m3-1)/2,(m5-2)/2,m4/2); else K582:=0:fi:

>        if  (m1-1)>=0  and  (m5-2)>=0  and  (m4-1)>=0  and  type((m1-
1)/2,integer)  and  type(m2/2,integer)  and  type((m5-2)/2,integer)
and    type((m4-1)/2,    integer)    and    type(m3/2,integer)    then
```

```
K592:=combinat[multinomial](4,2,0,1,1,0)*combinat[multinomial](n-
2,(m1-1)/2,m2/2,(m5-2)/2,(m4-1)/2,m3/2); else K592:=0:fi:

> if   (m2-1)>=0   and   (m3-1)>=0   and   (m4-2)>=0   and type((m2-
1)/2,integer)  and  type(m1/2,integer)  and  type((m3-1)/2,integer)
and   type((m4-2)/2,   integer)   and   type(m5/2,integer)   then
K602:=combinat[multinomial](4,2,0,1,1,0)*combinat[multinomial](n-
2,(m2-1)/2,m1/2,(m3-1)/2,(m4-2)/2,m5/2); else K602:=0:fi:

>          if   (m2-1)>=0   and   (m3-1)>=0   and   (m5-2)>=0   and type((m2-
1)/2,integer)     and     type((m3-1)/2,integer)     and     type((m5-
2)/2,integer) and type(m1/2, integer) and type(m4/2,integer) then
K612:=combinat[multinomial](4,2,1,0,0,1)*combinat[multinomial](n-
2,(m2-1)/2,(m3-1)/2,(m5-2)/2,m4/2,m1/2); else K612:=0:fi:

>          if   (m2-1)>=0   and   (m4-1)>=0   and   (m5-2)>=0   and type((m2-
1)/2,integer)  and  type(m1/2,integer)  and  type((m4-1)/2,integer)
and   type((m5-2)/2,   integer)   and   type(m3/2,integer)   then
K622:=combinat[multinomial](4,2,0,1,1,0)*combinat[multinomial](n-
2,(m2-1)/2,m1/2,(m4-1)/2,(m5-2)/2,m3/2); else K622:=0:fi:

>          if   (m3-1)>=0   and   (m5-2)>=0   and   (m4-1)>=0   and type((m3-
1)/2,integer)  and  type(m2/2,integer)  and  type((m5-2)/2,integer)
and   type((m4-1)/2,   integer)   and   type(m1/2,integer)   then
K632:=combinat[multinomial](4,2,0,1,1,0)*combinat[multinomial](n-
2,(m3-1)/2,m2/2,(m5-2)/2,(m4-1)/2,m1/2); else K632:=0:fi:

> if   (m1-1)>=0   and   (m2-1)>=0   and   (m3-1)>=0   and   (m4-1)>=0   and
type((m1-1)/2,integer)   and   type((m2-1)/2,integer)   and   type((m3-
1)/2,integer)  and  type((m4-1)/2,  integer)  and  type(m5/2,integer)
then
K642:=combinat[multinomial](4,1,1,1,1,0)*combinat[multinomial](n-
2,(m1-1)/2,(m2-1)/2,(m3-1)/2,(m4-1)/2,m5/2); else K642:=0:fi:

> if   (m1-1)>=0   and   (m2-1)>=0   and   (m3-1)>=0   and   (m5-1)>=0   and
type((m1-1)/2,integer)   and   type((m2-1)/2,integer)   and   type((m3-
1)/2,integer)  and  type((m5-1)/2,integer)  and  type(m4/2,integer)
then
K652:=combinat[multinomial](4,1,1,1,1,0)*combinat[multinomial](n-
2,(m1-1)/2,(m2-1)/2,(m3-1)/2,(m5-1)/2,m4/2); else K652:=0:fi:

> if   (m1-1)>=0   and   (m2-1)>=0   and   (m4-1)>=0   and   (m5-1)>=0   and
type((m1-1)/2,integer)   and   type((m2-1)/2,integer)   and   type((m4-
1)/2,integer)  and  type((m5-1)/2,integer)  and  type(m3/2,integer)
then
K662:=combinat[multinomial](4,1,1,1,1,0)*combinat[multinomial](n-
2,(m1-1)/2,(m2-1)/2,(m4-1)/2,(m5-1)/2,m3/2); else K662:=0:fi:

> if   (m1-1)>=0   and   (m3-1)>=0   and   (m4-1)>=0   and   (m5-1)>=0   and
type((m1-1)/2,integer)   and   type((m3-1)/2,integer)   and   type((m4-
1)/2,integer)  and  type((m5-1)/2,integer)  and  type(m2/2,integer)
then
K672:=combinat[multinomial](4,1,1,1,1,0)*combinat[multinomial](n-
2,(m1-1)/2,(m3-1)/2,(m4-1)/2,(m5-1)/2,m2/2); else K672:=0:fi:
```

```
> if  (m2-1)>=0  and  (m3-1)>=0  and  (m4-1)>=0  and  (m5-1)>=0  and
type((m2-1)/2,integer)   and   type((m3-1)/2,integer)   and   type((m4-
1)/2,integer)   and   type((m5-1)/2,integer)   and   type(m1/2,integer)
then
K682:=combinat[multinomial](4,1,1,1,1,0)*combinat[multinomial](n-
2,(m2-1)/2,(m3-1)/2,(m4-1)/2,(m5-1)/2,m1/2); else K682:=0:fi:

>
KT2:=K12+K22+K32+K42+K52+K62+K72+K82+K92+K102+K112+K122+K132+K142+
K152+K162+K172+K182+K192+K202+K212+K222+K232+K242+K252+K262+K272+K
282+K292+K302+K312+K322+K332+K342+K352+K362+K372+K382+K392+K402+K4
12+K422+K432+K442+K452+K462+K472+K482+K492+K502+K512+K522+K532+K54
2+K552+K562+K572+K582+K592+K602+K612+K622+K632+K642+K652+K662+K672
+K682:

>            if   type(m1/2,integer)   and   type(m2/2,integer)   and
type(m3/2,integer)   and   type(m4/2,integer)   and   type(m5/2,integer)
then    P:=combinat[multinomial](n,m1/2,m2/2,m3/2,m4/2,m5/2);    else
P:=0: fi:

> Ac(n,m1,m2,m3,m4,m5):=(1/(4*n))*(S1+((n/2)-1)*P)-(1/8)*KT2:

> Aa(n,m1,m2,m3,m4,m5):=(1/(2*n))*(S2+((n/2)+2)*P)+(1/4)*KT2:

> print (Ac(n,m1,m2,m3,m4,m5),Aa(n,m1,m2,m3,m4,m5));

> fi;

> end:

> end:

>
```

TO RUN THE PROGRAM: Type stereo(n,m1,m2,m3,m4,m5); and press Enter to get the number of chiral and achiral skeletons of CnXm1Ym2Zm3Um4Vm5 respectively.

Exemple:

```
> stereo(5,2,2,2,2,2);
```
$$5664, 72$$

```
> stereo(6,3,3,3,2,1);
```
$$46128, 144$$

```
> stereo(100,40,40,40,40,40);
```
5452595840540770602536623819044975098560266062192924184838792122 81   \
    9138887636244875514334998654788715162787623058028870925226865 74   \
    8000, 9134248905880772324781310060354332011170136339039982206811   \
    52063920

97

MAPLE PROGRAM FOR THE COMPUTATION OF THE NUMBERS OF CHIRAL AND ACHIRAL SKELETONS OF HETEROPOLYSUBSTITUTED MONOCYCLIC CYCLOALKANES CnXm1Ym2Zm3Um4Vm5Wm6

n = ring size, mi = number of non isomerizable substituents, i=1,2,3,4,5,6.

By Alphonse EMADAK

```
>
> restart:
```

Subprogram: To compute of the numbers of permutations induced by proper and improper symmetries of Dnh

```
> Ad1:=proc(x1)
> local y1;
> if type(x1,odd) or x1=2 then y1:=numtheory[phi](x1)
>               else   if   type(x1/2,odd)   and   x1<>2   then
y1:=2*numtheory[phi](x1)+numtheory[phi](x1/2);
>           else y1:=2*numtheory[phi](x1)
>       fi;
> fi;
> end:
```

Subprogram: To compute of the numbers of permutations induced by proper symmetries of Dnh

```
> Ad2:=proc(x2)
> local y2;
> if type(x2,odd) or type(x2,even) then y2:=numtheory[phi](x2);
> fi;
> end:
```

Subprogram: To compute the number Ac and Aa respectively of chiral and achiral graphs of heteropolysubstituted skeletons of cycloalkanes

```
>              stereo:=proc(n,m1,m2,m3,m4,m5,m6)              local
Dn,Dm1,Dm2,Dm3,Dm4,Dm5,Dm6,Dc,S1,S2,P,KT1,KT2,K11,K21,K31,K41,K51,
K61,K71,K81,K91,K101,K111,K121,K131,K141,K151,K161,K171,K181,K191,
K201,K211,K221,K12,K22,K32,K42,K52,K62,K72,K82,K92,K102,K112,K122,
K132,K142,K152,K162,K172,K182,K192,K202,K212,K222,K232,K242,K252,K
262,K272,K282,K292,K302,K312,K322,K332,K342,K352,K362,K372,K382,K3
92,K402,K412,K422,K432,K442,K452,K462,K472,K482,K492,K502,K512,K52
2,K532,K542,K552,K562,K572,K582,K592,K602,K612,K622,K632,K642,K652
,K662,K672,K682,K692,K702,K712,K722,K732,K742,K752,K762,K772,K782,
K792,K802,K812,K822,K832,K842,K852,K862,K872,K882,K892,K902,K912,K
922,K932,K942,K952,K962,K972,K982,K992,K1002,K1012,K1022,K1032,K10
```

98

42,K1052,K1062,K1072,K1082,K1092,K1102,K1112,K1122,K1132,K1142,K11
52,K1162,K1172,K1182,K1192,K1202,K1212,K1222,K1232:
CASE n ODD

```
>      if      type(n,odd)      then      Dn:=numtheory[divisors](2*n);
Dm1:=numtheory[divisors](m1):          Dm2:=numtheory[divisors](m2):
Dm3:=numtheory[divisors](m3):Dm4:=numtheory[divisors](m4):Dm5:=num
theory[divisors](m5):Dm6:=numtheory[divisors](m6):

> Dc:=  convert((Dn  intersect  Dm1  intersect  Dm2  intersect  Dm3
intersect Dm4 intersect Dm5 intersect Dm6) minus {2}, list):

>                                        S1:=sum('(2*Ad2(Dc[i])-
Ad1(Dc[i]))*combinat[multinomial](2*n/Dc[i],m1/Dc[i],m2/Dc[i],m3/D
c[i],m4/Dc[i],m5/Dc[i],m6/Dc[i])','i'=1..nops(Dc)):

>                                        S2:=sum('(Ad1(Dc[i])-
Ad2(Dc[i]))*combinat[multinomial](2*n/Dc[i],m1/Dc[i],m2/Dc[i],m3/D
c[i],m4/Dc[i],m5/Dc[i],m6/Dc[i])','i'=1..nops(Dc)):

>    if (m1-2)>=0 and type((m1-2)/2,integer) and type(m2/2,integer)
and      type(m3/2,integer)      and      type(m4/2,integer)      and
type(m5/2,integer)          and          type(m6/2,integer)          then
K11:=combinat[multinomial](2,2,0,0,0,0,0)*combinat[multinomial](n-
1,(m1-2)/2,m2/2,m3/2,m4/2,m5/2,m6/2): else K11:=0:fi:

>    if (m2-2)>=0 and type(m1/2,integer) and type((m2-2)/2,integer)
and      type(m3/2,integer)      and      type(m4/2,integer)      and
type(m6/2,integer)                                              then
K21:=combinat[multinomial](2,0,2,0,0,0,0)*combinat[multinomial](n-
1,m1/2,(m2-2)/2,m3/2,m4/2,m5/2,m6/2): else K21:=0:fi:

>     if (m3-2)>=0 and  type(m1/2,integer) and type(m2/2,integer)
and     type((m3-2)/2,integer)      and      type(m4/2,integer)      and
type(m5/2,integer)          and          type(m6/2,integer)          then
K31:=combinat[multinomial](2,0,0,2,0,0,0)*combinat[multinomial](n-
1,m1/2,m2/2,(m3-2)/2, m4/2,m5/2,m6/2): else K31:=0:fi:

>    if (m4-2)>=0 and  type(m1/2,integer) and type(m2/2,integer)
and      type(m3/2,integer)      and      type((m4-2)/2,integer)      and
type(m5/2,integer)          and          type(m6/2,integer)          then
K41:=combinat[multinomial](2,0,0,0,2,0,0)*combinat[multinomial](n-
1,m1/2,m2/2,(m3-2)/2, m4/2,m5/2,m6/2): else K41:=0:fi:

>    if (m5-2)>=0 and  type(m1/2,integer) and type(m2/2,integer)
and      type(m3/2,integer)      and      type((m5-2)/2,integer)      and
type(m4/2,integer)          and          type(m6/2,integer)          then
K51:=combinat[multinomial](2,0,0,0,0,2,0)*combinat[multinomial](n-
1,m1/2,m2/2,(m5-2)/2, m4/2,m3/2): else K51:=0:fi:

>    if (m6-2)>=0 and  type(m1/2,integer) and type(m2/2,integer)
and      type(m3/2,integer)      and      type((m6-2)/2,integer)      and
type(m4/2,integer)          and          type(m5/2,integer)          then
K61:=combinat[multinomial](2,0,0,0,0,0,2)*combinat[multinomial](n-
1,m1/2,m2/2,(m6-2)/2, m4/2,m3/2,m5/2): else K61:=0:fi:
```

```
>      if (m1-1)>=0 and (m2-1)>=0 and type((m1-1)/2,integer) and
type((m2-1)/2,integer)         and         type(m3/2,integer)         and
type(m4/2,integer) and type(m5/2,integer) and type(m6/2,integer)
then
K71:=combinat[multinomial](2,1,1,0,0,0,0)*combinat[multinomial](n-
1,(m1-1)/2,(m2-1)/2,m3/2,m4/2,m5/2,m6/2): else K71:=0:fi:

>      if (m1-1)>=0 and (m3-1)>=0 and type((m1-1)/2,integer) and
type(m2/2,integer)         and         type((m3-1)/2,integer)         and
type(m4/2,integer) and type(m5/2,integer) and type(m6/2,integer)
then
K81:=combinat[multinomial](2,1,0,1,0,0,0)*combinat[multinomial](n-
1,(m1-1)/2,m2/2,(m3-1)/2,m4/2,m5/2,m6/2): else K81:=0:fi:

>      if (m2-1)>=0 and (m3-1)>=0 and type(m1/2,integer) and
type((m2-1)/2,integer)        and        type((m3-1)/2,integer)        and
type(m4/2,integer) and type(m5/2,integer) and type(m6/2,integer)
then
K91:=combinat[multinomial](2,0,1,1,0,0,0)*combinat[multinomial](n-
1,m1/2,(m2-1)/2,(m3-1)/2,m4/2,m5/2,m6/2): else K91:=0:fi:

>      if (m1-1)>=0 and (m4-1)>=0 and type((m1-1)/2,integer) and
type(m2/2,integer)         and         type((m4-1)/2,integer)         and
type(m3/2,integer) and type(m5/2,integer) and type(m6/2,integer)
then
K101:=combinat[multinomial](2,1,0,0,1,0,0)*combinat[multinomial](n
-1,(m1-1)/2,m2/2,m3/2,(m4-1)/2,m5/2,m6/2): else K101:=0:fi:

>      if (m2-1)>=0 and (m4-1)>=0 and type(m1/2,integer) and
type((m2-1)/2,integer)        and        type((m4-1)/2,integer)        and
type(m3/2,integer) and type(m5/2,integer) and type(m6/2,integer)
then
K111:=combinat[multinomial](2,0,1,0,1,0)*combinat[multinomial](n-
1,m1/2,(m2-1)/2,m3/2,(m4-1)/2,m5/2,m6/2): else K111:=0:fi:

>      if (m3-1)>=0 and (m4-1)>=0 and type(m1/2,integer) and
type((m3-1)/2,integer)        and        type((m4-1)/2,integer)        and
type(m2/2,integer) and type(m5/2,integer) and type(m6/2,integer)
then
K121:=combinat[multinomial](2,0,0,1,1,0,0)*combinat[multinomial](n
-1,m1/2,(m3-1)/2,m2/2,(m4-1)/2,m5/2,m6/2): else K121:=0:fi:

>      if (m1-1)>=0 and (m5-1)>=0 and type((m1-1)/2,integer) and
type(m3/2,integer) and type(m4/2,integer) and type(m2/2,integer)
and      type((m5-1)/2,integer)        and        type(m6/2,integer)        then
K131:=combinat[multinomial](2,1,0,0,0,1,0)*combinat[multinomial](n
-1,(m1-1)/2,m2/2,m3/2,m4/2,(m5-1)/2,m6/2): else K131:=0:fi:

>      if (m2-1)>=0 and (m5-1)>=0 and type(m1/2,integer) and
type(m3/2,integer)         and         type(m4/2,integer)         and         type((m2-
1)/2,integer) and type((m5-1)/2,integer) and type(m6/2,integer)
then
K141:=combinat[multinomial](2,0,1,0,0,1,0)*combinat[multinomial](n
-1,m1/2,(m2-1)/2,m3/2,m4/2,(m5-1)/2,m6/2): else K141:=0:fi:
```

```
>        if  (m3-1)>=0  and  (m5-1)>=0  and  type(m1/2,integer)  and
type((m3-1)/2,integer)        and        type(m4/2,integer)        and
type(m2/2,integer)        and        type((m5-1)/2,integer)        and
type(m6/2,integer)                                                then
K151:=combinat[multinomial](2,0,0,1,0,1,0)*combinat[multinomial](n
-1,m1/2,m2/2,(m3-1)/2,m4/2,(m5-1)/2,m6/2): else K151:=0:fi:

>        if  (m4-1)>=0  and  (m5-1)>=0  and  type((m4-1)/2,integer)  and
type(m3/2,integer)  and  type(m1/2,integer)  and  type(m2/2,integer)
and      type((m5-1)/2,integer)      and      type(m6/2,integer)      then
K161:=combinat[multinomial](2,1,0,0,0,1)*combinat[multinomial](n-
1,(m4-1)/2,m2/2,m3/2,m1/2,(m5-1)/2,m6/2): else K161:=0:fi:

> if  (m1-1)>=0  and  (m6-1)>=0  and  type((m1-1)/2,integer)  and
type(m3/2,integer)  and  type(m4/2,integer)  and  type(m2/2,integer)
and      type((m6-1)/2,integer)      and      type(m5/2,integer)      then
K171:=combinat[multinomial](2,1,0,0,0,0,1)*combinat[multinomial](n
-1,(m1-1)/2,m2/2,m3/2,m4/2,(m6-1)/2,m5/2): else K171:=0:fi:

>        if  (m2-1)>=0  and  (m6-1)>=0  and  type(m1/2,integer)  and
type(m3/2,integer)      and      type(m4/2,integer)      and      type((m2-
1)/2,integer)  and  type((m6-1)/2,integer)  and  type(m5/2,integer)
then
K181:=combinat[multinomial](2,0,1,0,0,0,1)*combinat[multinomial](n
-1,m1/2,(m2-1)/2,m3/2,m4/2,(m6-1)/2,m5/2): else K181:=0:fi:

>        if  (m3-1)>=0  and  (m6-1)>=0  and  type(m1/2,integer)  and
type((m3-1)/2,integer)        and        type(m4/2,integer)        and
type(m2/2,integer)        and        type((m6-1)/2,integer)        and
type(m5/2,integer)                                                then
K191:=combinat[multinomial](2,0,0,1,0,0,1)*combinat[multinomial](n
-1,m1/2,m2/2,(m3-1)/2,m4/2,(m6-1)/2,m5/2): else K191:=0:fi:

>        if  (m4-1)>=0  and  (m6-1)>=0  and  type((m4-1)/2,integer)  and
type(m3/2,integer)  and  type(m1/2,integer)  and  type(m2/2,integer)
and      type((m6-1)/2,integer)      and      type(m5/2,integer)      then
K201:=combinat[multinomial](2,0,0,0,1,0,1)*combinat[multinomial](n
-1,(m4-1)/2,m2/2,m3/2,m1/2,(m6-1)/2,m5/2): else K201:=0:fi:

>        if  (m5-1)>=0  and  (m6-1)>=0  and  type((m5-1)/2,integer)  and
type(m3/2,integer)  and  type(m1/2,integer)  and  type(m2/2,integer)
and      type((m6-1)/2,integer)      and      type(m4/2,integer)      then
K211:=combinat[multinomial](2,0,0,0,0,1,1)*combinat[multinomial](n
-1,(m5-1)/2,m2/2,m3/2,m1/2,(m6-1)/2,m4/2): else K211:=0:fi:

>
KT1:=K11+K21+K31+K41+K51+K61+K71+K81+K91+K101+K111+K121+K131+K141+
K151+K161+K171+K181+K191+K201+K211:

>          if   type(m1/2,integer)   and   type(m2/2,integer)   and
type(m3/2,integer)  and  type(m4/2,integer)  and  type(m5/2,integer)
and                    type(m6/2,integer)                       then
P:=combinat[multinomial](n,m1/2,m2/2,m3/2,m4/2,m5/2,m6/2):     else
P:=0: fi:
```

101

```
> Ac(n,m1,m2,m3,m4,m5,m6):=(1/(4*n))*(S1+(n-1)*P)-(1/4)*KT1:

> Aa(n,m1,m2,m3,m4,m5,m6):=(1/(2*n))*(S2+P)+(1/2)*KT1:

> print (Ac(n,m1,m2,m3,m4,m5,m6),Aa(n,m1,m2,m3,m4,m5,m6));

> else                              # CASE n EVEN

>     if     type(n,even)      then     Dn:=numtheory[divisors](n);
Dm1:=numtheory[divisors](m1):          Dm2:=numtheory[divisors](m2):
Dm3:=numtheory[divisors](m3):Dm4:=numtheory[divisors](m4):Dm5:=num
theory[divisors](m5):Dm6:=numtheory[divisors](m6):

> Dc:= convert((Dn intersect Dm1 intersect Dm2 intersect Dm3
intersect Dm4 intersect Dm5 intersect Dm6) minus {2}, list):

>                                   S1:=sum('(2*Ad2(Dc[i])-
Ad1(Dc[i]))*combinat[multinomial](2*n/Dc[i],m1/Dc[i],m2/Dc[i],m3/D
c[i],m4/Dc[i],m5/Dc[i],m6/Dc[i])','i'=1..nops(Dc)):

>                                   S2:=sum('(Ad1(Dc[i])-
Ad2(Dc[i]))*combinat[multinomial](2*n/Dc[i],m1/Dc[i],m2/Dc[i],m3/D
c[i],m4/Dc[i],m5/Dc[i],m6/Dc[i])','i'=1..nops(Dc)):

>     if (m1-4)>=0 and type((m1-4)/2,integer) and type(m2/2,integer)
and      type(m3/2,integer)     and      type(m4/2,integer)      and
type(m5/2,integer)       and       type(m6/2,integer)        then
K12:=combinat[multinomial](4,4,0,0,0,0)*combinat[multinomial](n-
2,(m1-4)/2,m2/2,m3/2,m4/2,m5/2,m6/2); else K12:=0:fi:

>     if (m2-4)>=0 and type(m1/2,integer) and type((m2-4)/2,integer)
and      type(m3/2,integer)     and      type(m4/2,integer)      and
type(m5/2,integer)       and       type(m6/2,integer)        then
K22:=combinat[multinomial](4,0,4,0,0,0,0)*combinat[multinomial](n-
2,m1/2,(m2-4)/2,m3/2,m4/2,m5/2,m6/2); else K22:=0:fi:

>     if (m3-4)>=0 and type(m1/2,integer) and type(m2/2,integer) and
type((m3-4)/2,integer)      and       type(m4/2,integer)        and
type(m5/2,integer)       and       type(m6/2,integer)        then
K32:=combinat[multinomial](4,0,0,4,0,0,0)*combinat[multinomial](n-
2,m1/2,m2/2,(m3-4)/2,m4/2,m5/2,m6/2); else K32:=0:fi:

>     if (m4-4)>=0 and type(m1/2,integer) and type(m2/2,integer) and
type(m3/2,integer)       and       type((m4-4)/2,integer)       and
type(m5/2,integer)       and       type(m6/2,integer)        then
K32:=combinat[multinomial](4,0,0,0,4,0,0)*combinat[multinomial](n-
2,m1/2,m2/2,m3/2,(m4-4)/2,m5/2,m6/2); else K32:=0:fi:

>     if (m5-4)>=0 and type(m1/2,integer) and type(m2/2,integer) and
type(m3/2,integer)      and      type(m4/2,integer)     and     type((m5-
1)/2,integer)       and       type(m6/2,integer)              then
K42:=combinat[multinomial](4,0,0,0,0,4,0)*combinat[multinomial](n-
2,m1/2,m2/2,m3/2,(m5-4)/2,m4/2,m6/2); else K42:=0:fi:

>     if (m6-4)>=0 and type(m1/2,integer) and type(m2/2,integer) and
type(m3/2,integer)      and      type(m4/2,integer)     and     type((m6-
1)/2,integer)       and       type(m5/2,integer)             then
```

```
K42:=combinat[multinomial](4,0,0,0,0,0,4)*combinat[multinomial](n-
2,m1/2,m2/2,m3/2,(m6-4)/2,m4/2,m5/2); else K42:=0:fi:

>       if  (m1-3)>=0 and  (m2-1)>=0 and  type((m1-3)/2,integer)  and
type((m2-1)/2,integer)        and        type(m3/2,integer)         and
type(m4/2,integer)  and  type(m5/2,integer)  and  type(m6/2,integer)
then
K52:=combinat[multinomial](4,3,1,0,0,0,0)*combinat[multinomial](n-
2,(m1-3)/2,(m2-1)/2,m3/2,m4/2,m5/2,m6/2); else K52:=0:fi:

>       if  (m2-3)>=0  and  (m3-1)>=0  and  type(m1/2,integer)  and
type((m2-3)/2,integer)        and        type((m3-1)/2,integer)       and
type(m4/2,integer)  and  type(m5/2,integer)  and  type(m6/2,integer)
then
K62:=combinat[multinomial](4,0,3,1,0,0,0)*combinat[multinomial](n-
2,m1/2,(m2-3)/2,(m3-1)/2,m4/2,m5/2,m6/2); else K62:=0:fi:

>       if  (m1-3)>=0  and  (m3-1)>=0 and  type((m1-3)/2,integer)  and
type(m2/2,integer)        and        type((m3-1)/2,integer)          and
type(m4/2,integer)  and  type(m5/2,integer)  and  type(m6/2,integer)
then
K72:=combinat[multinomial](4,3,0,1,0,0,0)*combinat[multinomial](n-
2,(m1-3)/2,m2/2,(m3-1)/2,m4/2,m5/2,m6/2); else K72:=0:fi:

>       if  (m1-3)>=0  and  (m4-1)>=0 and  type((m1-3)/2,integer)  and
type(m2/2,integer)        and        type((m4-1)/2,integer)          and
type(m3/2,integer)  and  type(m5/2,integer)  and  type(m6/2,integer)
then
K82:=combinat[multinomial](4,3,0,0,1,0,0)*combinat[multinomial](n-
2,(m1-3)/2,m2/2,m3/2,(m4-1)/2,m5/2,m6/2); else K82:=0:fi:

>         if  (m2-3)>=0  and  (m4-1)>=0  and  type(m1/2,integer)  and
type((m2-3)/2,integer)        and        type((m4-1)/2,integer)       and
type(m3/2,integer)  and  type(m5/2,integer)  and  type(m6/2,integer)
then
K92:=combinat[multinomial](4,0,3,0,1,0,0)*combinat[multinomial](n-
2,m1/2,(m2-3)/2,m3/2,(m4-1)/2,m5/2,m6/2); else K92:=0:fi:

>         if  (m3-3)>=0  and  (m4-1)>=0  and  type(m1/2,integer)  and
type((m3-3)/2,integer)        and        type((m4-1)/2,integer)       and
type(m2/2,integer)  and  type(m5/2,integer)  and  type(m6/2,integer)
then
K102:=combinat[multinomial](4,0,0,3,1,0,0)*combinat[multinomial](n
-2,m1/2,m2/2,(m3-3)/2,(m4-1)/2,m5/2,m6/2); else K102:=0:fi:

>       if  (m1-3)>=0 and  (m5-1)>=0  and  type((m1-3)/2,integer)  and
type(m2/2,integer)  and  type(m3/2,integer)  and  type(m4/2,integer)
and      type((m5-1)/2,integer)      and      type(m6/2,integer)      then
K112:=combinat[multinomial](4,3,0,0,0,1,0)*combinat[multinomial](n
-2,(m1-3)/2,m2/2,m3/2,m4/2,(m5-1)/2,m6/2); else K112:=0:fi:

>       if  (m2-3)>=0  and  (m5-1)>=0 and  type((m2-3)/2,integer)  and
type(m1/2,integer)  and  type(m3/2,integer)  and  type(m4/2,integer)
and      type((m5-1)/2,integer)      and      type(m6/2,integer)      then
```

```
K122:=combinat[multinomial](4,3,0,1,0,0,0)*combinat[multinomial](n
-2,(m2-3)/2,m1/2,m3/2,m4/2,(m5-1)/2,m6/2); else K122:=0:fi:
>      if (m3-3)>=0 and (m5-1)>=0 and type((m3-3)/2,integer) and
type(m2/2,integer) and type(m1/2,integer) and type(m4/2,integer)
and     type((m5-1)/2,integer) and        type(m6/2,integer)    then
K132:=combinat[multinomial](4,3,0,1,0,0,0)*combinat[multinomial](n
-2,(m3-3)/2,m2/2,m1/2,m4/2,(m5-1)/2,m6/2); else K132:=0:fi:
>      if (m4-3)>=0 and (m5-1)>=0 and type((m4-3)/2,integer) and
type(m2/2,integer) and type(m3/2,integer) and type(m1/2,integer)
and     type((m5-1)/2,integer) and        type(m6/2,integer)    then
K142:=combinat[multinomial](4,3,0,1,0,0,0)*combinat[multinomial](n
-2,(m4-3)/2,m2/2,m3/2,m1/2,(m5-1)/2,m6/2); else K142:=0:fi:
>      if (m1-3)>=0 and (m6-1)>=0 and type((m1-3)/2,integer) and
type(m2/2,integer) and type(m3/2,integer) and type(m4/2,integer)
and     type((m6-1)/2,integer) and        type(m5/2,integer)    then
K152:=combinat[multinomial](4,3,0,0,0,0,1)*combinat[multinomial](n
-2,(m1-3)/2,m2/2,m3/2,m4/2,(m6-1)/2,m5/2); else K152:=0:fi:
>      if (m2-3)>=0 and (m6-1)>=0 and type((m2-3)/2,integer) and
type(m1/2,integer) and type(m3/2,integer) and type(m4/2,integer)
and     type((m6-1)/2,integer) and        type(m5/2,integer)    then
K162:=combinat[multinomial](4,3,0,1,0,0,0)*combinat[multinomial](n
-2,(m2-3)/2,m1/2,m3/2,m4/2,(m6-1)/2,m5/2); else K162:=0:fi:
>      if (m3-3)>=0 and (m5-1)>=0 and type((m3-3)/2,integer) and
type(m2/2,integer) and type(m1/2,integer) and type(m4/2,integer)
and     type((m5-1)/2,integer) and        type(m6/2,integer)    then
K172:=combinat[multinomial](4,0,0,3,0,1,0)*combinat[multinomial](n
-2,(m3-3)/2,m2/2,m1/2,m4/2,(m5-1)/2,m6/2); else K172:=0:fi:
>      if (m4-3)>=0 and (m6-1)>=0 and type((m4-3)/2,integer) and
type(m2/2,integer) and type(m3/2,integer) and type(m1/2,integer)
and     type((m6-1)/2,integer) and        type(m5/2,integer)    then
K182:=combinat[multinomial](4,0,0,0,3,0,1)*combinat[multinomial](n
-2,(m4-3)/2,m2/2,m3/2,m1/2,(m6-1)/2,m5/2); else K182:=0:fi:
>      if (m5-3)>=0 and (m6-1)>=0 and type((m5-3)/2,integer) and
type(m2/2,integer) and type(m3/2,integer) and type(m1/2,integer)
and     type((m6-1)/2,integer) and        type(m4/2,integer)    then
K192:=combinat[multinomial](4,0,0,0,0,3,1)*combinat[multinomial](n
-2,(m5-3)/2,m2/2,m3/2,m1/2,(m6-1)/2,m4/2); else K192:=0:fi:
>      if (m1-1)>=0 and (m2-3)>=0 and type((m1-1)/2,integer) and
type((m2-3)/2,integer)      and        type(m3/2,integer)       and
type(m4/2,integer) and type(m5/2,integer) and type(m6/2,integer)
then
K202:=combinat[multinomial](4,1,3,0,0,0,0)*combinat[multinomial](n
-2,(m1-1)/2,(m2-3)/2,m3/2,m4/2,m5/2,m6/2); else K202:=0:fi:
>      if (m2-1)>=0 and (m3-3)>=0 and type(m1/2,integer) and
type((m2-1)/2,integer)      and        type((m3-3)/2,integer)    and
type(m4/2,integer) and type(m5/2,integer) and type(m6/2,integer)
```

```
then
K212:=combinat[multinomial](4,0,1,3,0,0,0)*combinat[multinomial](n
-2,m1/2,(m2-1)/2,(m3-3)/2,m4/2,m5/2,m6/2); else K212:=0:fi:

>       if  (m1-1)>=0  and  (m3-3)>=0  and  type((m1-1)/2,integer)  and
type(m2/2,integer)        and        type((m3-3)/2,integer)        and
type(m4/2,integer)  and  type(m5/2,integer)  and  type(m6/2,integer)
then
K222:=combinat[multinomial](4,1,0,3,0,0,0)*combinat[multinomial](n
-2,(m1-1)/2,m2/2,(m3-3)/2,m4/2,m5/2,m6/2); else K222:=0:fi:

>       if  (m1-1)>=0  and  (m4-3)>=0  and  type((m1-1)/2,integer)  and
type(m2/2,integer)        and        type((m4-3)/2,integer)        and
type(m3/2,integer)  and  type(m5/2,integer)  and  type(m6/2,integer)
then
K232:=combinat[multinomial](4,1,0,0,3,0,0)*combinat[multinomial](n
-2,(m1-1)/2,m2/2,m3/2,(m4-3)/2,m5/2,m6/2); else K232:=0:fi:

>        if  (m2-1)>=0  and  (m4-3)>=0  and  type(m1/2,integer)  and
type((m2-1)/2,integer)      and      type((m4-3)/2,integer)        and
type(m3/2,integer)  and  type(m5/2,integer)  and  type(m6/2,integer)
then
K242:=combinat[multinomial](4,0,1,0,3,0,0)*combinat[multinomial](n
-2,m1/2,(m2-1)/2,m3/2,(m4-3)/2,m5/2,m6/2); else K242:=0:fi:

>        if  (m3-1)>=0  and  (m4-3)>=0  and  type(m1/2,integer)  and
type((m3-1)/2,integer)      and      type((m4-3)/2,integer)        and
type(m2/2,integer)  and  type(m5/2,integer)  and  type(m6/2,integer)
then
K252:=combinat[multinomial](4,0,0,1,3,0,0)*combinat[multinomial](n
-2,m1/2,m2/2,(m3-1)/2,(m4-3)/2,m5/2,m6/2); else K252:=0:fi:

>      if  (m1-1)>=0  and  (m5-3)>=0  and  type((m1-1)/2,integer)  and
type(m2/2,integer)  and  type(m3/2,integer)  and  type(m4/2,integer)
and      type((m5-3)/2,integer)      and      type(m6/2,integer)      then
K262:=combinat[multinomial](4,1,0,0,0,3,0)*combinat[multinomial](n
-2,(m1-1)/2,m2/2,m3/2,m4/2,(m5-3)/2,m6/2); else K262:=0:fi:

>       if  (m2-1)>=0  and  (m5-3)>=0  and  type((m2-1)/2,integer)  and
type(m1/2,integer)  and  type(m3/2,integer)  and  type(m4/2,integer)
and      type((m5-3)/2,integer)      and      type(m6/2,integer)      then
K272:=combinat[multinomial](4,0,1,0,0,3,0)*combinat[multinomial](n
-2,(m2-1)/2,m1/2,m3/2,m4/2,(m5-3)/2,m6/2); else K272:=0:fi:

>       if  (m3-1)>=0  and  (m5-3)>=0  and  type((m3-1)/2,integer)  and
type(m2/2,integer)  and  type(m1/2,integer)  and  type(m4/2,integer)
and      type((m5-3)/2,integer)      and      type(m6/2,integer)      then
K282:=combinat[multinomial](4,0,0,1,0,3,0)*combinat[multinomial](n
-2,(m3-1)/2,m2/2,m1/2,m4/2,(m5-3)/2,m6/2); else K282:=0:fi:

>     if  (m4-1)>=0  and  (m5-3)>=0  and  type((m4-1)/2,integer)  and
type(m2/2,integer)  and  type(m3/2,integer)  and  type(m1/2,integer)
and      type((m5-3)/2,integer)      and      type(m6/2,integer)      then
K292:=combinat[multinomial](4,0,0,0,1,3,0)*combinat[multinomial](n
```

```
-2,(m4-1)/2,m2/2,m3/2,m1/2,(m5-3)/2,m6/2); else K292:=0:fi:

>     if (m1-1)>=0 and (m6-3)>=0 and type((m1-1)/2,integer) and
type(m2/2,integer) and type(m3/2,integer) and type(m4/2,integer)
and    type((m6-3)/2,integer)    and    type(m5/2,integer)    then
K302:=combinat[multinomial](4,3,0,0,0,0,1)*combinat[multinomial](n
-2,(m1-1)/2,m2/2,m3/2,m4/2,(m6-3)/2,m5/2); else K302:=0:fi:

>     if (m2-1)>=0 and (m6-3)>=0 and type((m2-1)/2,integer) and
type(m1/2,integer) and type(m3/2,integer) and type(m4/2,integer)
and    type((m6-3)/2,integer)    and    type(m5/2,integer)    then
K312:=combinat[multinomial](4,3,0,1,0,0,0)*combinat[multinomial](n
-2,(m2-1)/2,m1/2,m3/2,m4/2,(m6-3)/2,m5/2); else K312:=0:fi:

>     if (m3-1)>=0 and (m5-3)>=0 and type((m3-1)/2,integer) and
type(m2/2,integer) and type(m1/2,integer) and type(m4/2,integer)
and    type((m5-3)/2,integer)    and    type(m6/2,integer)    then
K322:=combinat[multinomial](4,0,0,1,0,3,0)*combinat[multinomial](n
-2,(m3-1)/2,m2/2,m1/2,m4/2,(m5-3)/2,m6/2); else K322:=0:fi:

>     if (m4-1)>=0 and (m6-3)>=0 and type((m4-1)/2,integer) and
type(m2/2,integer) and type(m3/2,integer) and type(m1/2,integer)
and    type((m6-3)/2,integer)    and    type(m5/2,integer)    then
K332:=combinat[multinomial](4,0,0,0,1,0,3)*combinat[multinomial](n
-2,(m4-1)/2,m2/2,m3/2,m1/2,(m6-3)/2,m5/2); else K332:=0:fi:

>     if (m5-1)>=0 and (m6-3)>=0 and type((m5-1)/2,integer) and
type(m2/2,integer) and type(m3/2,integer) and type(m1/2,integer)
and    type((m6-3)/2,integer)    and    type(m4/2,integer)    then
K342:=combinat[multinomial](4,0,0,0,0,1,3)*combinat[multinomial](n
-2,(m5-1)/2,m2/2,m3/2,m1/2,(m6-3)/2,m4/2); else K342:=0:fi:

>     if (m1-2)>=0 and (m2-2)>=0 and type((m1-2)/2,integer) and
type((m2-2)/2,integer)      and      type(m3/2,integer)      and
type(m4/2,integer) and type(m5/2,integer) and type(m6/2,integer)
then
K352:=combinat[multinomial](4,2,2,0,0,0,0)*combinat[multinomial](n
-2,(m1-2)/2,(m2-2)/2,m3/2,m4/2,m5/2,m6/2); else K352:=0:fi:

>      if (m2-2)>=0 and (m3-2)>=0 and type(m1/2,integer) and
type((m2-2)/2,integer)      and      type((m3-2)/2,integer)      and
type(m4/2,integer) and type(m5/2,integer) and type(m6/2,integer)
then
K362:=combinat[multinomial](4,0,2,2,0,0,0)*combinat[multinomial](n
-2,m1/2,(m2-2)/2,(m3-2)/2,m4/2,m5/2,m6/2); else K362:=0:fi:

>     if (m1-2)>=0 and (m3-2)>=0 and type((m1-2)/2,integer) and
type(m2/2,integer)      and      type((m3-2)/2,integer)      and
type(m4/2,integer)      and      type(m6/2,integer)      then
K372:=combinat[multinomial](4,2,0,2,0,0)*combinat[multinomial](n-
2,(m1-2)/2,m2/2,(m3-2)/2,m4/2,m5/2,m6/2); else K372:=0:fi:

>      if (m3-2)>=0 and (m4-2)>=0 and type(m1/2,integer) and
type(m2/2,integer)    and    type((m3-2)/2,integer)    and    type((m4-
2)/2,integer) and type(m5/2,integer) and type(m6/2,integer) then
```

106

```
K382:=combinat[multinomial](4,0,0,2,2,0,0)*combinat[multinomial](n
-2,m1/2,m2/2,(m3-2)/2,(m4-2)/2,m5/2,m6/2); else K382:=0:fi:

>       if (m1-2)>=0 and (m4-2)>=0 and type((m1-2)/2,integer) and
type(m2/2,integer)    and    type(m3/2,integer)    and    type((m4-
2)/2,integer) and type(m5/2,integer) and type(m6/2,integer) then
K392:=combinat[multinomial](4,2,0,0,2,0,0)*combinat[multinomial](n
-2,(m1-2)/2,m2/2,m3/2,(m4-2)/2,m5/2,m6/2); else K392:=0:fi:

>       if (m2-2)>=0 and (m4-2)>=0 and type(m1/2,integer) and
type((m2-2)/2,integer)     and     type(m3/2,integer)     and
type(m4/2,integer) and type(m5/2,integer) and type(m6/2,integer)
then
K402:=combinat[multinomial](4,0,2,0,2,0,0)*combinat[multinomial](n
-2,m1/2,(m2-2)/2,m3/2,(m4-2)/2,m5/2,m6/2); else K402:=0:fi:

>       if (m1-2)>=0 and (m5-2)>=0 and type((m1-2)/2,integer) and
type(m2/2,integer)    and    type(m3/2,integer)    and    type((m5-
2)/2,integer) and type(m4/2,integer) and type(m6/2,integer) then
K412:=combinat[multinomial](4,2,0,0,0,2,0)*combinat[multinomial](n
-2,(m1-2)/2,m2/2,m3/2,m4/2,(m5-2)/2,m6/2); else K412:=0:fi:

>       if (m2-2)>=0 and (m5-2)>=0 and type((m2-2)/2,integer) and
type(m1/2,integer)    and    type(m3/2,integer)    and    type((m5-
2)/2,integer) and type(m4/2,integer) and type(m6/2,integer) then
K412:=combinat[multinomial](4,0,2,0,0,2,0)*combinat[multinomial](n
-2,(m2-2)/2,m1/2,m3/2,m4/2,(m5-2)/2,m6/2); else K412:=0:fi:

>       if (m3-2)>=0 and (m5-2)>=0 and type((m3-2)/2,integer) and
type(m1/2,integer)    and    type(m2/2,integer)    and    type((m5-
2)/2,integer) and type(m4/2,integer) and type(m6/2,integer) then
K422:=combinat[multinomial](4,0,0,2,0,2,0)*combinat[multinomial](n
-2,(m3-2)/2,m1/2,m2/2,m4/2,(m5-2)/2,m6/2); else K422:=0:fi:

> if    (m4-2)>=0    and    (m5-2)>=0    and    type((m4-2)/2,integer)    and
type(m1/2,integer)    and    type(m3/2,integer)    and    type((m5-
2)/2,integer) and type(m2/2,integer) and type(m6/2,integer) then
K432:=combinat[multinomial](4,0,0,0,2,2)*combinat[multinomial](n-
2,(m4-2)/2,m1/2,m3/2,m2/2,(m5-2)/2,m6/2); else K432:=0:fi:

>       if (m1-2)>=0 and (m6-2)>=0 and type((m1-2)/2,integer) and
type(m2/2,integer)    and    type(m3/2,integer)    and    type((m6-
2)/2,integer) and type(m4/2,integer) and type(m5/2,integer) then
K442:=combinat[multinomial](4,2,0,0,0,2,0)*combinat[multinomial](n
-2,(m1-2)/2,m2/2,m3/2,m4/2,(m6-2)/2,m5/2); else K442:=0:fi:

>       if (m2-2)>=0 and (m6-2)>=0 and type((m2-2)/2,integer) and
type(m1/2,integer)    and    type(m3/2,integer)    and    type((m6-
2)/2,integer) and type(m4/2,integer) and type(m5/2,integer) then
K452:=combinat[multinomial](4,0,2,0,0,2,0)*combinat[multinomial](n
-2,(m2-2)/2,m1/2,m3/2,m4/2,(m6-2)/2,m5/2); else K452:=0:fi:

>       if (m3-2)>=0 and (m6-2)>=0 and type((m3-2)/2,integer) and
type(m1/2,integer)    and    type(m2/2,integer)    and    type((m6-
2)/2,integer) and type(m4/2,integer) and type(m5/2,integer) then
```

```
K462:=combinat[multinomial](4,0,0,2,0,2,0)*combinat[multinomial](n
-2,(m3-2)/2,m1/2,m2/2,m4/2,(m6-2)/2,m5/2); else K462:=0:fi:

>       if  (m4-2)>=0  and  (m6-2)>=0  and  type((m4-2)/2,integer)  and
type(m1/2,integer)     and     type(m3/2,integer)     and     type((m6-
2)/2,integer) and type(m2/2,integer) and type(m5/2,integer)   then
K472:=combinat[multinomial](4,0,0,0,2,2)*combinat[multinomial](n-
2,(m4-2)/2,m1/2,m3/2,m2/2,(m6-2)/2,m5/2); else K472:=0:fi:

>       if  (m5-2)>=0  and  (m6-2)>=0  and  type((m5-2)/2,integer)  and
type(m2/2,integer)     and     type(m3/2,integer)     and     type((m6-
2)/2,integer) and type(m4/2,integer) and type(m1/2,integer)   then
K482:=combinat[multinomial](4,2,0,0,0,2,0)*combinat[multinomial](n
-2,(m5-2)/2,m2/2,m3/2,m4/2,(m6-2)/2,m1/2); else K482:=0:fi:

>       if  (m1-2)>=0  and  (m2-1)>=0  and  (m3-1)>=0  and  type((m1-
2)/2,integer)     and     type((m2-1)/2,integer)     and     type((m3-
1)/2,integer)  and type(m4/2, integer) and type(m5/2,integer)  and
type(m6/2,integer)                                              then
K492:=combinat[multinomial](4,2,1,1,0,0,0)*combinat[multinomial](n
-2,(m1-2)/2,(m2-1)/2,(m3-1)/2,m4/2,m5/2,m6/2); else K492:=0:fi:

>       if  (m1-2)>=0  and  (m2-1)>=0  and  (m4-1)>=0  and  type((m1-
2)/2,integer)  and  type((m2-1)/2,integer)  and  type(m3/2,integer)
and    type((m4-1)/2,    integer)    and    type(m5/2,integer)    and
type(m6/2,integer)                                              then
K502:=combinat[multinomial](4,2,1,0,1,0,0)*combinat[multinomial](n
-2,(m1-2)/2,(m2-1)/2,m3/2,(m4-1)/2,m5/2,m6/2); else K502:=0:fi:

>       if  (m1-2)>=0  and  (m3-1)>=0  and  (m4-1)>=0  and  type((m1-
2)/2,integer)  and  type(m2/2,integer)  and  type((m3-1)/2,integer)
and    type((m4-1)/2,    integer)    and    type(m5/2,integer)    and
type(m6/2,integer)                                              then
K512:=combinat[multinomial](4,2,0,1,1,0,0)*combinat[multinomial](n
-2,(m1-2)/2,m2/2,(m3-1)/2,(m4-1)/2,m5/2,m6/2); else K512:=0:fi:

>       if  (m1-2)>=0  and  (m2-1)>=0  and  (m5-1)>=0  and  type((m1-
2)/2,integer)     and     type((m2-1)/2,integer)     and     type((m5-
1)/2,integer) and type(m4/2, integer) and type(m3/2,integer) and
type(m6/2,integer)                                              then
K522:=combinat[multinomial](4,2,1,0,0,1,0)*combinat[multinomial](n
-2,(m1-2)/2,(m2-1)/2,(m5-1)/2,m4/2,m3/2,m6/2); else K522:=0:fi:

>       if  (m1-2)>=0  and  (m3-1)>=0  and  (m5-1)>=0  and  type((m1-
2)/2,integer)  and  type(m2/2,integer)  and  type((m3-1)/2,integer)
and    type((m5-1)/2,    integer)    and    type(m4/2,integer)    and
type(m6/2,integer)                                              then
K532:=combinat[multinomial](4,2,0,1,1,0)*combinat[multinomial](n-
2,(m1-2)/2,m2/2,(m3-1)/2,(m5-1)/2,m4/2,m6/2); else K532:=0:fi:

>       if  (m1-2)>=0  and  (m5-1)>=0  and  (m4-1)>=0  and  type((m1-
2)/2,integer)  and  type(m2/2,integer)  and  type((m5-1)/2,integer)
and    type((m4-1)/2,    integer)    and    type(m3/2,integer)    and
type(m6/2,integer)                                              then
```

```
K542:=combinat[multinomial](4,2,0,1,1,0,0)*combinat[multinomial](n
-2,(m1-2)/2,m2/2,(m5-1)/2,(m4-1)/2,m3/2,m6/2); else K542:=0:fi:

>       if (m1-2)>=0 and (m2-1)>=0 and (m6-1)>=0 and type((m1-
2)/2,integer) and type((m2-1)/2,integer) and type(m3/2,integer)
and   type((m6-1)/2,   integer)   and   type(m5/2,integer)   and
type(m4/2,integer)                                          then
K552:=combinat[multinomial](4,2,1,0,1,0,0)*combinat[multinomial](n
-2,(m1-2)/2,(m2-1)/2,m3/2,(m6-1)/2,m5/2,m4/2); else K552:=0:fi:

>       if (m1-2)>=0 and (m3-1)>=0 and (m6-1)>=0 and type((m1-
2)/2,integer) and type(m2/2,integer) and type((m3-1)/2,integer)
and   type((m6-1)/2,   integer)   and   type(m5/2,integer)   and
type(m4/2,integer)                                          then
K562:=combinat[multinomial](4,2,0,1,1,0,0)*combinat[multinomial](n
-2,(m1-2)/2,m2/2,(m3-1)/2,(m6-1)/2,m5/2,m4/2); else K562:=0:fi:

>       if (m1-2)>=0 and (m6-1)>=0 and (m4-1)>=0 and type((m1-
2)/2,integer) and type(m2/2,integer) and type((m6-1)/2,integer)
and   type((m4-1)/2,   integer)   and   type(m3/2,integer)   and
type(m5/2,integer)                                          then
K572:=combinat[multinomial](4,2,0,1,1,0,0)*combinat[multinomial](n
-2,(m1-2)/2,m2/2,(m6-1)/2,(m4-1)/2,m3/2,m5/2); else K572:=0:fi:

>       if (m1-2)>=0 and (m5-1)>=0 and (m6-1)>=0 and type((m1-
2)/2,integer) and type(m2/2,integer) and type((m5-1)/2,integer)
and   type((m6-1)/2,   integer)   and   type(m3/2,integer)   and
type(m4/2,integer)                                          then
K582:=combinat[multinomial](4,2,0,1,1,0,0)*combinat[multinomial](n
-2,(m1-2)/2,m2/2,(m5-1)/2,(m6-1)/2,m3/2,m4/2); else K582:=0:fi:

>       if (m2-2)>=0 and (m3-1)>=0 and (m4-1)>=0 and type((m2-
2)/2,integer) and type(m1/2,integer) and type((m3-1)/2,integer)
and   type((m4-1)/2,   integer)   and   type(m5/2,integer)and
type(m6/2,integer)                                          then
K592:=combinat[multinomial](4,0,2,1,1,0,0)*combinat[multinomial](n
-2,(m2-2)/2,m1/2,(m3-1)/2,(m4-1)/2,m5/2,m6/2); else K592:=0:fi:

>       if (m2-2)>=0 and (m3-1)>=0 and (m5-1)>=0 and type((m2-
2)/2,integer) and type(m6/2,integer) and type((m3-1)/2,integer)
and   type((m5-1)/2,integer)   and   type(m1/2,   integer)   and
type(m4/2,integer)                                          then
K602:=combinat[multinomial](4,2,1,0,0,1,0)*combinat[multinomial](n
-2,(m2-2)/2,(m3-1)/2,(m5-1)/2,m4/2,m1/2,m6/2); else K602:=0:fi:

>       if (m2-2)>=0 and (m4-1)>=0 and (m5-1)>=0 and type((m2-
2)/2,integer) and type(m1/2,integer) and type((m4-1)/2,integer)
and   type((m5-1)/2,   integer)   and   type(m3/2,integer)   and
type(m6/2,integer)                                          then
K612:=combinat[multinomial](4,2,0,1,1,0,0)*combinat[multinomial](n
-2,(m2-2)/2,m1/2,(m4-1)/2,(m5-1)/2,m3/2,m6/2); else K612:=0:fi:

>       if (m3-2)>=0 and (m5-1)>=0 and (m4-1)>=0 and type((m3-
2)/2,integer) and type(m2/2,integer) and type((m5-1)/2,integer)
```

109

```
and    type((m4-1)/2,    integer)    and    type(m1/2,integer)    and
type(m6/2,integer)                                                then
K622:=combinat[multinomial](4,2,0,1,1,0,0)*combinat[multinomial](n
-2,(m3-2)/2,m2/2,(m5-1)/2,(m4-1)/2,m1/2,m6/2); else K622:=0:fi:
>        if  (m2-2)>=0  and  (m3-1)>=0  and  (m6-1)>=0  and  type((m2-
2)/2,integer)  and  type(m5/2,integer)  and  type((m3-1)/2,integer)
and    type((m6-1)/2,integer)    and    type(m1/2,    integer)    and
type(m4/2,integer)                                                then
K632:=combinat[multinomial](4,2,1,0,0,1,0)*combinat[multinomial](n
-2,(m2-2)/2,(m3-1)/2,(m6-1)/2,m4/2,m1/2,m5/2); else K632:=0:fi:
>        if  (m2-2)>=0  and  (m4-1)>=0  and  (m6-1)>=0  and  type((m2-
2)/2,integer)  and  type(m1/2,integer)  and  type((m4-1)/2,integer)
and    type((m6-1)/2,    integer)    and    type(m3/2,integer)    and
type(m5/2,integer)                                                then
K642:=combinat[multinomial](4,2,0,1,1,0,0)*combinat[multinomial](n
-2,(m2-2)/2,m1/2,(m4-1)/2,(m6-1)/2,m3/2,m5/2); else K642:=0:fi:
>        if  (m2-2)>=0  and  (m5-1)>=0  and  (m6-1)>=0  and  type((m2-
2)/2,integer)  and  type(m1/2,integer)  and  type((m5-1)/2,integer)
and    type((m6-1)/2,    integer)    and    type(m3/2,integer)    and
type(m4/2,integer)                                                then
K652:=combinat[multinomial](4,2,0,1,1,0,0)*combinat[multinomial](n
-2,(m2-2)/2,m1/2,(m5-1)/2,(m6-1)/2,m3/2,m4/2); else K652:=0:fi:
>        if  (m3-2)>=0  and  (m6-1)>=0  and  (m4-1)>=0  and  type((m3-
2)/2,integer)  and  type(m2/2,integer)  and  type((m6-1)/2,integer)
and    type((m4-1)/2,    integer)    and    type(m1/2,integer)    and
type(m5/2,integer)                                                then
K662:=combinat[multinomial](4,2,0,1,1,0,0)*combinat[multinomial](n
-2,(m3-2)/2,m2/2,(m6-1)/2,(m4-1)/2,m1/2,m5/2); else K662:=0:fi:
>        if  (m3-2)>=0  and  (m5-1)>=0  and  (m6-1)>=0  and  type((m3-
2)/2,integer)  and  type(m2/2,integer)  and  type((m5-1)/2,integer)
and    type((m6-1)/2,    integer)    and    type(m1/2,integer)    and
type(m4/2,integer)                                                then
K672:=combinat[multinomial](4,2,0,1,1,0,0)*combinat[multinomial](n
-2,(m3-2)/2,m2/2,(m5-1)/2,(m6-1)/2,m1/2,m4/2); else K672:=0:fi:
>        if  (m4-2)>=0  and  (m5-1)>=0  and  (m6-1)>=0  and  type((m4-
2)/2,integer)  and  type(m2/2,integer)  and  type((m5-1)/2,integer)
and    type((m6-1)/2,    integer)    and    type(m1/2,integer)    and
type(m3/2,integer)                                                then
K682:=combinat[multinomial](4,2,0,1,1,0,0)*combinat[multinomial](n
-2,(m4-2)/2,m2/2,(m5-1)/2,(m6-1)/2,m1/2,m3/2); else K682:=0:fi:
>        if  (m1-1)>=0  and  (m2-2)>=0  and  (m3-1)>=0  and  type((m1-
1)/2,integer)    and    type((m2-2)/2,integer)    and    type((m3-
1)/2,integer)  and  type(m4/2,  integer)  and  type(m5/2,integer)  and
type(m6/2,integer)                                                then
K692:=combinat[multinomial](4,2,1,1,0,0,0)*combinat[multinomial](n
-2,(m1-1)/2,(m2-2)/2,(m3-1)/2,m4/2,m5/2,m6/2); else K692:=0:fi:
```

110

```
>        if  (m1-1)>=0  and  (m2-2)>=0  and  (m4-1)>=0  and  type((m1-
1)/2,integer)  and  type((m2-2)/2,integer)  and  type(m3/2,integer)
and    type((m4-1)/2,    integer)    and    type(m5/2,integer)    and
type(m6/2,integer)                                              then
K702:=combinat[multinomial](4,2,1,0,1,0,0)*combinat[multinomial](n
-2,(m1-1)/2,(m2-2)/2,m3/2,(m4-1)/2,m5/2,m6/2); else K702:=0:fi:

>        if  (m1-1)>=0  and  (m3-2)>=0  and  (m4-1)>=0  and  type((m1-
1)/2,integer)  and  type(m2/2,integer)  and  type((m3-2)/2,integer)
and    type((m4-1)/2,    integer)    and    type(m5/2,integer)    and
type(m6/2,integer)                                              then
K712:=combinat[multinomial](4,2,0,1,1,0,0)*combinat[multinomial](n
-2,(m1-1)/2,m2/2,(m3-2)/2,(m4-1)/2,m5/2,m6/2); else K712:=0:fi:

>        if  (m1-1)>=0  and  (m2-2)>=0  and  (m5-1)>=0  and  type((m1-
1)/2,integer)       and       type((m2-2)/2,integer)       and       type((m5-
1)/2,integer)  and  type(m4/2,  integer)  and  type(m3/2,integer)  and
type(m6/2,integer)                                              then
K722:=combinat[multinomial](4,2,1,0,0,1,0)*combinat[multinomial](n
-2,(m1-1)/2,(m2-2)/2,(m5-1)/2,m4/2,m3/2,m6/2); else K722:=0:fi:

>        if  (m1-1)>=0  and  (m3-2)>=0  and  (m5-1)>=0  and  type((m1-
1)/2,integer)  and  type(m2/2,integer)  and  type((m3-2)/2,integer)
and    type((m5-1)/2,    integer)    and    type(m4/2,integer)    and
type(m6/2,integer)                                              then
K732:=combinat[multinomial](4,2,0,1,1,0)*combinat[multinomial](n-
2,(m1-1)/2,m2/2,(m3-2)/2,(m5-1)/2,m4/2,m6/2); else K732:=0:fi:

>        if  (m1-1)>=0  and  (m5-1)>=0  and  (m4-2)>=0  and  type((m1-
1)/2,integer)  and  type(m2/2,integer)  and  type((m5-1)/2,integer)
and    type((m4-2)/2,    integer)    and    type(m3/2,integer)    and
type(m6/2,integer)                                              then
K742:=combinat[multinomial](4,2,0,1,1,0,0)*combinat[multinomial](n
-2,(m1-1)/2,m2/2,(m5-1)/2,(m4-2)/2,m3/2,m6/2); else K742:=0:fi:

>        if  (m1-1)>=0  and  (m2-2)>=0  and  (m6-1)>=0  and  type((m1-
1)/2,integer)  and  type((m2-2)/2,integer)  and  type(m3/2,integer)
and    type((m6-1)/2,    integer)    and    type(m5/2,integer)    and
type(m4/2,integer)                                              then
K752:=combinat[multinomial](4,2,1,0,1,0,0)*combinat[multinomial](n
-2,(m1-1)/2,(m2-2)/2,m3/2,(m6-1)/2,m5/2,m4/2); else K752:=0:fi:

>        if  (m1-1)>=0  and  (m3-2)>=0  and  (m6-1)>=0  and  type((m1-
1)/2,integer)  and  type(m2/2,integer)  and  type((m3-2)/2,integer)
and    type((m6-1)/2,    integer)    and    type(m5/2,integer)    and
type(m4/2,integer)                                              then
K762:=combinat[multinomial](4,2,0,1,1,0,0)*combinat[multinomial](n
-2,(m1-1)/2,m2/2,(m3-2)/2,(m6-1)/2,m5/2,m4/2); else K762:=0:fi:

>        if  (m1-1)>=0  and  (m6-1)>=0  and  (m4-2)>=0  and  type((m1-
1)/2,integer)  and  type(m2/2,integer)  and  type((m6-1)/2,integer)
and    type((m4-2)/2,    integer)    and    type(m3/2,integer)    and
type(m5/2,integer)                                              then
K772:=combinat[multinomial](4,2,0,1,1,0,0)*combinat[multinomial](n
```

```
-2,(m1-1)/2,m2/2,(m6-1)/2,(m4-2)/2,m3/2,m5/2); else K772:=0:fi:

>        if  (m1-1)>=0  and  (m5-2)>=0  and  (m6-1)>=0  and  type((m1-
1)/2,integer)  and  type(m2/2,integer)  and  type((m5-2)/2,integer)
and    type((m6-1)/2,    integer)    and    type(m3/2,integer)    and
type(m4/2,integer)                                               then
K782:=combinat[multinomial](4,2,0,1,1,0,0)*combinat[multinomial](n
-2,(m1-1)/2,m2/2,(m5-2)/2,(m6-1)/2,m3/2,m4/2); else K782:=0:fi:

>        if  (m2-1)>=0  and  (m3-2)>=0  and  (m4-1)>=0  and  type((m2-
1)/2,integer)  and  type(m1/2,integer)  and  type((m3-2)/2,integer)
and    type((m4-1)/2,    integer)    and    type(m5/2,integer)and
type(m6/2,integer)                                               then
K792:=combinat[multinomial](4,0,2,1,1,0,0)*combinat[multinomial](n
-2,(m1-2)/2,m1/2,(m3-2)/2,(m4-1)/2,m5/2,m6/2); else K792:=0:fi:

>        if  (m2-1)>=0  and  (m3-2)>=0  and  (m5-1)>=0  and  type((m2-
1)/2,integer)  and  type(m6/2,integer)  and  type((m3-2)/2,integer)
and    type((m5-1)/2,integer)    and    type(m1/2,    integer)    and
type(m4/2,integer)                                               then
K802:=combinat[multinomial](4,2,1,0,0,1,0)*combinat[multinomial](n
-2,(m2-1)/2,(m3-2)/2,(m5-1)/2,m4/2,m1/2,m6/2); else K802:=0:fi:

>        if  (m2-1)>=0  and  (m4-2)>=0  and  (m5-1)>=0  and  type((m2-
1)/2,integer)  and  type(m1/2,integer)  and  type((m4-2)/2,integer)
and    type((m5-1)/2,    integer)    and    type(m3/2,integer)    and
type(m6/2,integer)                                               then
K812:=combinat[multinomial](4,2,0,1,1,0,0)*combinat[multinomial](n
-2,(m2-1)/2,m1/2,(m4-2)/2,(m5-1)/2,m3/2,m6/2); else K812:=0:fi:

>        if  (m3-1)>=0  and  (m5-1)>=0  and  (m4-2)>=0  and  type((m3-
1)/2,integer)  and  type(m2/2,integer)  and  type((m5-1)/2,integer)
and    type((m4-2)/2,    integer)    and    type(m1/2,integer)    and
type(m6/2,integer)                                               then
K822:=combinat[multinomial](4,2,0,1,1,0,0)*combinat[multinomial](n
-2,(m3-1)/2,m2/2,(m5-1)/2,(m4-2)/2,m1/2,m6/2); else K822:=0:fi:

>        if  (m2-1)>=0  and  (m3-2)>=0  and  (m6-1)>=0  and  type((m2-
1)/2,integer)  and  type(m5/2,integer)  and  type((m3-2)/2,integer)
and    type((m6-1)/2,integer)    and    type(m1/2,    integer)    and
type(m4/2,integer)                                               then
K832:=combinat[multinomial](4,2,1,0,0,1,0)*combinat[multinomial](n
-2,(m2-1)/2,(m3-2)/2,(m6-1)/2,m4/2,m1/2,m5/2); else K832:=0:fi:

>        if  (m2-1)>=0  and  (m4-2)>=0  and  (m6-1)>=0  and  type((m2-
1)/2,integer)  and  type(m1/2,integer)  and  type((m4-2)/2,integer)
and    type((m6-1)/2,    integer)    and    type(m3/2,integer)    and
type(m5/2,integer)                                               then
K842:=combinat[multinomial](4,2,0,1,1,0,0)*combinat[multinomial](n
-2,(m2-1)/2,m1/2,(m4-2)/2,(m6-1)/2,m3/2,m5/2); else K842:=0:fi:

>     if  (m2-1)>=0  and  (m5-2)>=0  and  (m6-1)>=0  and  type((m2-
1)/2,integer)  and  type(m1/2,integer)  and  type((m5-2)/2,integer)
and    type((m6-1)/2,    integer)    and    type(m3/2,integer)    and
```

112

```
type(m4/2,integer)                                                  then
K852:=combinat[multinomial](4,2,0,1,1,0,0)*combinat[multinomial](n
-2,(m2-1)/2,m1/2,(m5-2)/2,(m6-1)/2,m3/2,m4/2); else K852:=0:fi:

>        if (m3-1)>=0 and (m6-1)>=0 and (m4-2)>=0 and type((m3-
1)/2,integer) and type(m2/2,integer) and type((m6-1)/2,integer)
and   type((m4-2)/2,   integer)   and   type(m1/2,integer)   and
type(m5/2,integer)                                                  then
K862:=combinat[multinomial](4,2,0,1,1,0,0)*combinat[multinomial](n
-2,(m3-1)/2,m2/2,(m6-1)/2,(m4-2)/2,m1/2,m5/2); else K862:=0:fi:

>        if (m3-1)>=0 and (m5-2)>=0 and (m6-1)>=0 and type((m3-
1)/2,integer) and type(m2/2,integer) and type((m5-2)/2,integer)
and   type((m6-1)/2,   integer)   and   type(m1/2,integer)   and
type(m4/2,integer)                                                  then
K872:=combinat[multinomial](4,2,0,1,1,0,0)*combinat[multinomial](n
-2,(m3-1)/2,m2/2,(m5-2)/2,(m6-1)/2,m1/2,m4/2); else K872:=0:fi:

>        if (m4-1)>=0 and (m5-2)>=0 and (m6-1)>=0 and type((m4-
1)/2,integer) and type(m2/2,integer) and type((m5-2)/2,integer)
and   type((m6-1)/2,   integer)   and   type(m1/2,integer)   and
type(m3/2,integer)                                                  then
K882:=combinat[multinomial](4,2,0,1,1,0,0)*combinat[multinomial](n
-2,(m4-1)/2,m2/2,(m5-2)/2,(m6-1)/2,m1/2,m3/2); else K882:=0:fi:

>        if (m1-1)>=0 and (m2-1)>=0 and (m3-2)>=0 and type((m1-
1)/2,integer)     and     type((m2-1)/2,integer)     and     type((m3-
2)/2,integer) and type(m4/2, integer) and type(m5/2,integer) and
type(m6/2,integer)                                                  then
K892:=combinat[multinomial](4,2,1,1,0,0,0)*combinat[multinomial](n
-2,(m1-1)/2,(m2-1)/2,(m3-2)/2,m4/2,m5/2,m6/2); else K892:=0:fi:

>        if (m1-1)>=0 and (m2-1)>=0 and (m4-2)>=0 and type((m1-
1)/2,integer)  and  type((m2-1)/2,integer)  and  type(m3/2,integer)
and   type((m4-2)/2,   integer)   and   type(m5/2,integer)   and
type(m6/2,integer)                                                  then
K902:=combinat[multinomial](4,2,1,0,1,0,0)*combinat[multinomial](n
-2,(m1-1)/2,(m2-1)/2,m3/2,(m4-2)/2,m5/2,m6/2); else K902:=0:fi:

>        if (m1-1)>=0 and (m3-1)>=0 and (m4-2)>=0 and type((m1-
1)/2,integer)  and  type(m2/2,integer)  and  type((m3-1)/2,integer)
and   type((m4-2)/2,   integer)   and   type(m5/2,integer)   and
type(m6/2,integer)                                                  then
K912:=combinat[multinomial](4,2,0,1,1,0,0)*combinat[multinomial](n
-2,(m1-1)/2,m2/2,(m3-1)/2,(m4-2)/2,m5/2,m6/2); else K912:=0:fi:

>        if (m1-1)>=0 and (m2-1)>=0 and (m5-2)>=0 and type((m1-
1)/2,integer)     and     type((m2-1)/2,integer)     and     type((m5-
2)/2,integer) and type(m4/2, integer) and type(m3/2,integer) and
type(m6/2,integer)                                                  then
K922:=combinat[multinomial](4,2,1,0,0,1,0)*combinat[multinomial](n
-2,(m1-1)/2,(m2-1)/2,(m5-2)/2,m4/2,m3/2,m6/2); else K922:=0:fi:

>        if (m1-1)>=0 and (m3-1)>=0 and (m5-2)>=0 and type((m1-
```

113

```
1)/2,integer)   and   type(m2/2,integer)   and   type((m3-1)/2,integer)
and     type((m5-2)/2,     integer)     and     type(m4/2,integer)     and
type(m6/2,integer)                                                     then
K932:=combinat[multinomial](4,2,0,1,1,0)*combinat[multinomial](n-
2,(m1-1)/2,m2/2,(m3-1)/2,(m5-2)/2,m4/2,m6/2); else K932:=0:fi:

>       if  (m1-1)>=0  and  (m5-2)>=0  and  (m4-1)>=0  and  type((m1-
1)/2,integer)   and   type(m2/2,integer)   and   type((m5-2)/2,integer)
and     type((m4-1)/2,     integer)     and     type(m3/2,integer)     and
type(m6/2,integer)                                                     then
K942:=combinat[multinomial](4,2,0,1,1,0,0)*combinat[multinomial](n
-2,(m1-1)/2,m2/2,(m5-2)/2,(m4-1)/2,m3/2,m6/2); else K942:=0:fi:

>       if  (m1-1)>=0  and  (m2-1)>=0  and  (m6-2)>=0  and  type((m1-
1)/2,integer)   and   type((m2-1)/2,integer)   and   type(m3/2,integer)
and     type((m6-2)/2,     integer)     and     type(m5/2,integer)     and
type(m4/2,integer)                                                     then
K952:=combinat[multinomial](4,2,1,0,1,0,0)*combinat[multinomial](n
-2,(m1-1)/2,(m2-1)/2,m3/2,(m6-2)/2,m5/2,m4/2); else K952:=0:fi:

>       if  (m1-1)>=0  and  (m3-1)>=0  and  (m6-2)>=0  and  type((m1-
1)/2,integer)   and   type(m2/2,integer)   and   type((m3-1)/2,integer)
and     type((m6-2)/2,     integer)     and     type(m5/2,integer)     and
type(m4/2,integer)                                                     then
K962:=combinat[multinomial](4,2,0,1,1,0,0)*combinat[multinomial](n
-2,(m1-1)/2,m2/2,(m3-1)/2,(m6-2)/2,m5/2,m4/2); else K962:=0:fi:

>       if  (m1-1)>=0  and  (m6-2)>=0  and  (m4-1)>=0  and  type((m1-
1)/2,integer)   and   type(m2/2,integer)   and   type((m6-2)/2,integer)
and     type((m4-1)/2,     integer)     and     type(m3/2,integer)     and
type(m5/2,integer)                                                     then
K972:=combinat[multinomial](4,2,0,1,1,0,0)*combinat[multinomial](n
-2,(m1-1)/2,m2/2,(m6-2)/2,(m4-1)/2,m3/2,m5/2); else K972:=0:fi:

>       if  (m1-1)>=0  and  (m5-1)>=0  and  (m6-2)>=0  and  type((m1-
1)/2,integer)   and   type(m2/2,integer)   and   type((m5-1)/2,integer)
and     type((m6-2)/2,     integer)     and     type(m3/2,integer)     and
type(m4/2,integer)                                                     then
K982:=combinat[multinomial](4,2,0,1,1,0,0)*combinat[multinomial](n
-2,(m1-1)/2,m2/2,(m5-1)/2,(m6-2)/2,m3/2,m4/2); else K982:=0:fi:

>       if  (m2-1)>=0  and  (m3-1)>=0  and  (m4-2)>=0  and  type((m2-
1)/2,integer)   and   type(m1/2,integer)   and   type((m3-1)/2,integer)
and     type((m4-2)/2,     integer)     and     type(m5/2,integer)and
type(m6/2,integer)                                                     then
K992:=combinat[multinomial](4,0,2,1,1,0,0)*combinat[multinomial](n
-2,(m1-2)/2,m1/2,(m3-1)/2,(m4-2)/2,m5/2,m6/2); else K992:=0:fi:

>       if  (m2-1)>=0  and  (m3-1)>=0  and  (m5-2)>=0  and  type((m2-
1)/2,integer)   and   type(m6/2,integer)   and   type((m3-1)/2,integer)
and     type((m5-2)/2,integer)     and     type(m1/2,     integer)     and
type(m4/2,integer)                                                     then
K1002:=combinat[multinomial](4,2,1,0,0,1,0)*combinat[multinomial](
n-2,(m2-1)/2,(m3-1)/2,(m5-2)/2,m4/2,m1/2,m6/2); else K1002:=0:fi:
```

```
>        if  (m2-1)>=0  and  (m4-1)>=0  and  (m5-2)>=0  and  type((m2-
1)/2,integer)  and  type(m1/2,integer)  and  type((m4-1)/2,integer)
and    type((m5-2)/2,    integer)    and    type(m3/2,integer)    and
type(m6/2,integer)                                              then
K1012:=combinat[multinomial](4,2,0,1,1,0,0)*combinat[multinomial](
n-2,(m2-1)/2,m1/2,(m4-1)/2,(m5-2)/2,m3/2,m6/2); else K1012:=0:fi:

>        if  (m3-1)>=0  and  (m5-2)>=0  and  (m4-1)>=0  and  type((m3-
1)/2,integer)  and  type(m2/2,integer)  and  type((m5-2)/2,integer)
and    type((m4-1)/2,    integer)    and    type(m1/2,integer)    and
type(m6/2,integer)                                              then
K1022:=combinat[multinomial](4,2,0,1,1,0,0)*combinat[multinomial](
n-2,(m3-1)/2,m2/2,(m5-2)/2,(m4-1)/2,m1/2,m6/2); else K1022:=0:fi:

>        if  (m2-1)>=0  and  (m3-1)>=0  and  (m6-2)>=0  and  type((m2-
1)/2,integer)  and  type(m5/2,integer)  and  type((m3-1)/2,integer)
and    type((m6-2)/2,integer)    and    type(m1/2,    integer)    and
type(m4/2,integer)                                              then
K1032:=combinat[multinomial](4,2,1,0,0,1,0)*combinat[multinomial](
n-2,(m2-1)/2,(m3-1)/2,(m6-2)/2,m4/2,m1/2,m5/2); else K1032:=0:fi:

>        if  (m2-1)>=0  and  (m4-1)>=0  and  (m6-2)>=0  and  type((m2-
1)/2,integer)  and  type(m1/2,integer)  and  type((m4-1)/2,integer)
and    type((m6-2)/2,    integer)    and    type(m3/2,integer)    and
type(m5/2,integer)                                              then
K1042:=combinat[multinomial](4,2,0,1,1,0,0)*combinat[multinomial](
n-2,(m2-1)/2,m1/2,(m4-1)/2,(m6-2)/2,m3/2,m5/2); else K1042:=0:fi:

>        if  (m2-1)>=0  and  (m5-1)>=0  and  (m6-2)>=0  and  type((m2-
1)/2,integer)  and  type(m1/2,integer)  and  type((m5-1)/2,integer)
and    type((m6-2)/2,    integer)    and    type(m3/2,integer)    and
type(m4/2,integer)                                              then
K1052:=combinat[multinomial](4,2,0,1,1,0,0)*combinat[multinomial](
n-2,(m2-1)/2,m1/2,(m5-1)/2,(m6-2)/2,m3/2,m4/2); else K1052:=0:fi:

>        if  (m3-1)>=0  and  (m6-2)>=0  and  (m4-1)>=0  and  type((m3-
1)/2,integer)  and  type(m2/2,integer)  and  type((m6-2)/2,integer)
and    type((m4-1)/2,    integer)    and    type(m1/2,integer)    and
type(m5/2,integer)                                              then
K1062:=combinat[multinomial](4,2,0,1,1,0,0)*combinat[multinomial](
n-2,(m3-1)/2,m2/2,(m6-2)/2,(m4-1)/2,m1/2,m5/2); else K1062:=0:fi:

>        if  (m3-1)>=0  and  (m5-1)>=0  and  (m6-2)>=0  and  type((m3-
1)/2,integer)  and  type(m2/2,integer)  and  type((m5-1)/2,integer)
and    type((m6-2)/2,    integer)    and    type(m1/2,integer)    and
type(m4/2,integer)                                              then
K1072:=combinat[multinomial](4,2,0,1,1,0,0)*combinat[multinomial](
n-2,(m3-1)/2,m2/2,(m5-1)/2,(m6-2)/2,m1/2,m4/2); else K1072:=0:fi:

>        if  (m4-1)>=0  and  (m5-1)>=0  and  (m6-2)>=0  and  type((m4-
1)/2,integer)  and  type(m2/2,integer)  and  type((m5-1)/2,integer)
and    type((m6-2)/2,    integer)    and    type(m1/2,integer)    and
type(m3/2,integer)                                              then
K1082:=combinat[multinomial](4,2,0,1,1,0,0)*combinat[multinomial](
```

```
n-2, (m4-1)/2,m2/2, (m5-1)/2, (m6-2)/2,m1/2,m3/2) ; else K1082:=0:fi:

>       if (m1-1)>=0 and (m2-1)>=0 and (m3-1)>=0 and (m4-1)>=0 and
type((m1-1)/2,integer) and type((m2-1)/2,integer) and type((m3-
1)/2,integer) and type((m4-1)/2, integer) and type(m5/2,integer)
and                  type(m6/2,integer)                  then
K1092:=combinat[multinomial](4,1,1,1,1,0,0)*combinat[multinomial](
n-2, (m1-1)/2, (m2-1)/2, (m3-1)/2, (m4-1)/2,m5/2,m6/2) ;       else
K1092:=0:fi:

>       if (m1-1)>=0 and (m2-1)>=0 and (m3-1)>=0 and (m5-1)>=0 and
type((m1-1)/2,integer) and type((m2-1)/2,integer) and type((m3-
1)/2,integer) and type((m5-1)/2,integer) and type(m4/2,integer)
and                  type(m6/2,integer)                  then
K1102:=combinat[multinomial](4,1,1,1,1,0,0)*combinat[multinomial](
n-2, (m1-1)/2, (m2-1)/2, (m3-1)/2, (m5-1)/2,m4/2,m6/2) ;       else
K1102:=0:fi:

>       if (m1-1)>=0 and (m2-1)>=0 and (m4-1)>=0 and (m5-1)>=0 and
type((m1-1)/2,integer) and type((m2-1)/2,integer) and type((m4-
1)/2,integer) and type((m5-1)/2,integer) and type(m3/2,integer)
and                  type(m6/2,integer)                  then
K1112:=combinat[multinomial](4,1,1,1,1,0,0)*combinat[multinomial](
n-2, (m1-1)/2, (m2-1)/2, (m4-1)/2, (m5-1)/2,m3/2,m6/2) ;       else
K1112:=0:fi:

>       if (m1-1)>=0 and (m3-1)>=0 and (m4-1)>=0 and (m5-1)>=0 and
type((m1-1)/2,integer) and type((m3-1)/2,integer) and type((m4-
1)/2,integer) and type((m5-1)/2,integer) and type(m2/2,integer)
and                  type(m6/2,integer)                  then
K1122:=combinat[multinomial](4,1,1,1,1,0,0)*combinat[multinomial](
n-2, (m1-1)/2, (m3-1)/2, (m4-1)/2, (m5-1)/2,m2/2,m6/2) ;       else
K1122:=0:fi:

>       if (m2-1)>=0 and (m3-1)>=0 and (m4-1)>=0 and (m5-1)>=0 and
type((m2-1)/2,integer) and type((m3-1)/2,integer) and type((m4-
1)/2,integer) and type((m5-1)/2,integer) and type(m1/2,integer)
and                  type(m6/2,integer)                  then
K1132:=combinat[multinomial](4,1,1,1,1,0,0)*combinat[multinomial](
n-2, (m2-1)/2, (m3-1)/2, (m4-1)/2, (m5-1)/2,m1/2,m6/2) ;       else
K1132:=0:fi:

>       if (m2-1)>=0 and (m3-1)>=0 and (m4-1)>=0 and (m6-1)>=0 and
type((m2-1)/2,integer) and type((m3-1)/2,integer) and type((m4-
1)/2,integer) and type((m6-1)/2,integer) and type(m1/2,integer)
and                  type(m5/2,integer)                  then
K1142:=combinat[multinomial](4,1,1,1,1,0,0)*combinat[multinomial](
n-2, (m2-1)/2, (m3-1)/2, (m4-1)/2, (m6-1)/2,m1/2,m5/2) ;       else
K1142:=0:fi:

>       if (m2-1)>=0 and (m3-1)>=0 and (m6-1)>=0 and (m5-1)>=0 and
type((m2-1)/2,integer) and type((m3-1)/2,integer) and type((m6-
1)/2,integer) and type((m5-1)/2,integer) and type(m1/2,integer)
and                  type(m4/2,integer)                  then
```

```
K1152:=combinat[multinomial](4,1,1,1,1,0,0)*combinat[multinomial](
n-2,(m2-1)/2,(m3-1)/2,(m5-1)/2,(m6-1)/2,m1/2,m4/2);              else
K1152:=0:fi:

>     if (m2-1)>=0 and (m5-1)>=0 and (m4-1)>=0 and (m6-1)>=0 and
type((m2-1)/2,integer) and type((m5-1)/2,integer) and type((m4-
1)/2,integer) and type((m6-1)/2,integer) and type(m1/2,integer)
and                     type(m3/2,integer)                      then
K1162:=combinat[multinomial](4,1,1,1,1,0,0)*combinat[multinomial](
n-2,(m2-1)/2,(m5-1)/2,(m4-1)/2,(m6-1)/2,m1/2,m3/2);              else
K1162:=0:fi:

>     if (m5-1)>=0 and (m3-1)>=0 and (m4-1)>=0 and (m6-1)>=0 and
type((m5-1)/2,integer) and type((m3-1)/2,integer) and type((m4-
1)/2,integer) and type((m6-1)/2,integer) and type(m1/2,integer)
and                     type(m2/2,integer)                      then
K1172:=combinat[multinomial](4,1,1,1,1,0,0)*combinat[multinomial](
n-2,(m5-1)/2,(m3-1)/2,(m4-1)/2,(m6-1)/2,m1/2,m2/2);              else
K1172:=0:fi:

>     if (m1-1)>=0 and (m2-1)>=0 and (m3-1)>=0 and (m6-1)>=0 and
type((m1-1)/2,integer) and type((m2-1)/2,integer) and type((m3-
1)/2,integer) and type((m6-1)/2, integer) and type(m5/2,integer)
and                     type(m4/2,integer)                      then
K1182:=combinat[multinomial](4,1,1,1,1,0,0)*combinat[multinomial](
n-2,(m1-1)/2,(m2-1)/2,(m3-1)/2,(m6-1)/2,m5/2,m4/2);              else
K1182:=0:fi:

>     if (m1-1)>=0 and (m2-1)>=0 and (m4-1)>=0 and (m6-1)>=0 and
type((m1-1)/2,integer) and type((m2-1)/2,integer) and type((m4-
1)/2,integer) and type((m6-1)/2, integer) and type(m5/2,integer)
and                     type(m3/2,integer)                      then
K1192:=combinat[multinomial](4,1,1,1,1,0,0)*combinat[multinomial](
n-2,(m1-1)/2,(m2-1)/2,(m4-1)/2,(m6-1)/2,m5/2,m3/2);              else
K1192:=0:fi:

>     if (m1-1)>=0 and (m2-1)>=0 and (m5-1)>=0 and (m6-1)>=0 and
type((m1-1)/2,integer) and type((m2-1)/2,integer) and type((m5-
1)/2,integer) and type((m6-1)/2, integer) and type(m3/2,integer)
and                     type(m4/2,integer)                      then
K1202:=combinat[multinomial](4,1,1,1,1,0,0)*combinat[multinomial](
n-2,(m1-1)/2,(m2-1)/2,(m5-1)/2,(m6-1)/2,m3/2,m4/2);              else
K1202:=0:fi:

>     if (m1-1)>=0 and (m4-1)>=0 and (m5-1)>=0 and (m6-1)>=0 and
type((m1-1)/2,integer) and type((m4-1)/2,integer) and type((m5-
1)/2,integer) and type((m6-1)/2, integer) and type(m3/2,integer)
and                     type(m2/2,integer)                      then
K1212:=combinat[multinomial](4,1,1,1,1,0,0)*combinat[multinomial](
n-2,(m1-1)/2,(m4-1)/2,(m5-1)/2,(m6-1)/2,m3/2,m2/2);              else
K1212:=0:fi:

>     if (m1-1)>=0 and (m3-1)>=0 and (m5-1)>=0 and (m6-1)>=0 and
type((m1-1)/2,integer) and type((m3-1)/2,integer) and type((m5-
```

117

```
1)/2,integer)  and  type((m6-1)/2,  integer)  and  type(m4/2,integer)
and                        type(m2/2,integer)                      then
K1222:=combinat[multinomial](4,1,1,1,1,0,0)*combinat[multinomial](
n-2,(m1-1)/2,(m3-1)/2,(m5-1)/2,(m6-1)/2,m4/2,m2/2);           else
K1222:=0:fi:

>       if  (m1-1)>=0 and  (m4-1)>=0 and  (m3-1)>=0 and  (m6-1)>=0 and
type((m1-1)/2,integer)   and  type((m4-1)/2,integer)  and  type((m3-
1)/2,integer)  and  type((m6-1)/2,  integer)  and  type(m5/2,integer)
and                        type(m2/2,integer)                      then
K1232:=combinat[multinomial](4,1,1,1,1,0,0)*combinat[multinomial](
n-2,(m1-1)/2,(m4-1)/2,(m3-1)/2,(m6-1)/2,m5/2,m2/2);           else
K1232:=0:fi:

>
KT2:=K12+K22+K32+K42+K52+K62+K72+K82+K92+K102+K112+K122+K132+K142+
K152+K162+K172+K182+K192+K202+K212+K222+K232+K242+K252+K262+K272+K
282+K292+K302+K312+K322+K332+K342+K352+K362+K372+K382+K392+K402+K4
12+K422+K432+K442+K452+K462+K472+K482+K492+K502+K512+K522+K532+K54
2+K552+K562+K572+K582+K592+K602+K612+K622+K632+K642+K652+K662+K672
+K682+K692+K702+K712+K722+K732+K742+K752+K762+K772+K782+K792+K802+
K812+K822+K832+K842+K852+K862+K872+K882+K892+K902+K912+K922+K932+K
942+K952+K962+K972+K982+K992+K1002+K1012+K1022+K1032+K1042+K1052+K
1062+K1072+K1082+K1092+K1102+K1112+K1122+K1132+K1142+K1152+K1162+K
1172+K1182+K1192+K1202+K1212+K1222+K1232:

>              if   type(m1/2,integer)   and   type(m2/2,integer)   and
type(m3/2,integer)  and  type(m4/2,integer)  and  type(m5/2,integer)
and                        type(m6/2,integer)                      then
P:=combinat[multinomial](n,m1/2,m2/2,m3/2,m4/2,m5/2,m6/2);    else
P:=0: fi:

> Ac(n,m1,m2,m3,m4,m5,m6):=(1/(4*n))*(S1+((n/2)-1)*P)-(1/8)*KT2:

> Aa(n,m1,m2,m3,m4,m5,m6):=(1/(2*n))*(S2+((n/2)+2)*P)+(1/4)*KT2:

> print (Ac(n,m1,m2,m3,m4,m5,m6),Aa(n,m1,m2,m3,m4,m5,m6));

> fi;

> end:

> end:
```

TO RUN THE PROGRAM: Type stereo(n,m1,m2,m3,m4,m5,m6); and press Enter to get the number of chiral and achiral skeletons of CnXm1Ym2Zm3Um4Vm5Wm6 respectively.

Exemple :

```
> stereo(6,2,2,2,1,3,2);
```
$$207744, 312$$

```
> stereo(8,4,3,5,2,1,1);
```
$$18918360, 1080$$

118

```
> stereo(100,40,40,40,40,20,20);
```
75162140967691545924757348037748162479987005624894483604545556412   \
    202254611417046972611451889787604040920917941808152390438683586   \
    783446902416000 , 177955828745070753420380293481508349835371411547   \
    37062560809802229214171840

```
> stereo(12,7,5,6,3,1,2);
```
                              2473653591360 , 302400

```
> stereo(25,12,12,5,7,11,3);
```
                          9151287371142830516794243852800 , 0

```
> stereo(4,1,1,2,1,2,1);
```
                                   624, 12

```
> stereo(6,2,2,2,2,2,2);
```
                                 311658, 804

```
> stereo(9,4,2,2,2,5,3);
```
                              1286480160 , 10080

```
> stereo(7,3,5,2,1,2,1);
```
                                 1081080, 0

```
> stereo(10,6,4,2,3,3,2);
```
                              24443180160 , 77280

```
> stereo(11,3,4,5,5,2,3);
```
                              1026615189600 , 0

```
> stereo(12,7,5,6,3,1,2);
```
                              2473653591360 , 302400

```
> stereo(13,7,5,6,3,3,2);
```
                              247365374256000 , 0

119

www.ingramcontent.com/pod-product-compliance
Lightning Source LLC
Chambersburg PA
CBHW021108210326
41598CB00017B/1379